# Automating Army Convoys

Technical and Tactical Risks and Opportunities

SHAWN MCKAY, MATTHEW E. BOYER, NAHOM M. BEYENE,
MICHAEL LERARIO, MATTHEW W. LEWIS, KARLYN D. STANLEY,
RANDALL STEEB, BRADLEY WILSON, KATHERYN GIGLIO

ARROYO CENTER

For more information on this publication, visit www.rand.org/t/RR2406

**Library of Congress Cataloging-in-Publication Data** is available for this publication.
ISBN: 978-1-9774-0039-0

Published by the RAND Corporation, Santa Monica, Calif.
© Copyright 2020 RAND Corporation
**RAND**® is a registered trademark.

www.rand.org

# Preface

This report documents research and analysis conducted as part of a project entitled "Implementation of Autonomous Vehicles in the CS & CSS Force Structure," sponsored by the Program Executive Office Combat Support and Combat Service Support. The purpose of the project was to identify and assess the force implications and risks posed by the anticipated near- to mid-term opportunities for automating Army convoy trucks. A minimally manned bridging option leading to the use of automated Army trucks is developed in this report to address the current technical and tactical risks of concepts requiring the use of unmanned, automated trucks in Army convoys. This report may be of interest to individuals and organizations planning for or currently pursuing autonomous vehicle technology.

RAND operates under a "Federal-Wide Assurance" (FWA00003425) and complies with the *Code of Federal Regulations for the Protection of Human Subjects Under United States Law* (45 CFR 46), also known as "the Common Rule," as well as with the implementation guidance set forth in DoD Instruction 3216.02. As applicable, this compliance includes reviews and approvals by RAND's Institutional Review Board (the Human Subjects Protection Committee) and by the U.S. Army. The views of sources utilized in this study are solely their own and do not represent the official policy or position of DoD or the U.S. government.

This research was conducted within RAND Arroyo Center's Forces and Logistics Program. RAND Arroyo Center, part of the

# Contents

# Figures and Tables

## Figures

# Tables

# Summary

The U.S. Army has thousands of ground vehicles and is interested in harnessing the potential benefits of emerging self-driving technology. In theory, automation could create efficiencies and save lives by reducing the number of personnel operating in combat zones. The use of automated trucks in convoys is of special interest: Recent combat operations have continually demonstrated the vulnerability of convoys due to their fundamental requirement for delivering sustainment supplies over long distances of unsecured routes. This operational reality of convoy missions makes them particularly vulnerable to attack and ambush.

Given that fully automated convoys are not yet feasible, the Army research and development communities have been testing automated truck concepts in which manned and unmanned vehicles perform cooperatively in convoy operations. These concepts are promising because they have the potential to reduce the number of soldiers needed in a convoy, but the technical and tactical feasibility of these concepts need further examination. It is not fully understood what kinds of technological and operational changes these concepts of using automated trucks in convoys will introduce. Thus, it is important that the Army carefully consider the state of the art and the potential changes this new technology may introduce in order to manage with foresight.

Understanding the need for full analysis, the Program Executive Office Combat Support and Combat Service Support (PEO CS&CSS) asked RAND Arroyo Center to assess the risks that automated truck acquisition may experience in development and wider Army operations.

This research aims to determine the specific risks and risk mitigations for the development of automated trucks in the near to mid-term future (one to five years). The research team developed two research areas and related questions to address this problem:

- **Technology:** How mature is autonomous vehicle (AV) technology for Army convoy operations? What are potential risks in deploying this technology through the medium term?
- **Doctrine, organization, training, materiel, leadership, personnel, facilities, and policies (DOTmLPF-P):**[1] What effects will automated convoys have on Army force structure, operation planning, and execution?

The team used multiple methods to address these questions, including subject-matter expert (SME) interviews, a review of Army and commercial test data, and sociotechnical systems (STeS) analysis. These efforts resulted in a new automation concept option for the Army to consider in the more immediate term, as well as several recommendations for moving ahead in development and utilization more generally.

## Three Automated Convoy Concepts for the Army to Consider

Many of the Army research and development activities in this arena have focused on a concept in which automated unmanned trucks follow the path of a manned truck in a convoy operation. We term this the *partially unmanned* (PU) employment concept. Because of some significant technical and tactical risks we discovered early in the study, we created a second concept for the Army to consider: the *minimally manned* (MM) employment concept.[2] This concept is being used by

---

[1]   The tactical assessment covers select aspects of DOTmLPF-P but is not a full DOTmLPF-P assessment. For this reason, we use the general term *tactical* instead.

[2]   We explicitly use the term *employment concept* instead of *concept* to distinguish that these ideas differ mainly in how the automated trucks are used operationally. The PU and MM

many companies developing automated vehicles today and should be considered as a bridging concept between today's human-operated convoys and the PU employment concept. The primary difference in the two concepts lies in the manning of the follower trucks. In the PU employment concept, the follower trucks are completely unmanned. In the MM employment concept, there is a single soldier in the driver's seat to monitor the automated system and driving environment, but there is no passenger, as there is in traditional convoy operations. The third concept is a longer-term science and technology vision in which all the cargo trucks in the convoy are unmanned. In this concept, the Army trucks will be fully autonomous, greatly reducing the soldiers needed during the convoy operation. We term this the *fully autonomous* (FA) employment concept. Analyzing these concepts gave more breadth to the analysis, but, more importantly, the MM concept offers the Army a feasible, and most likely necessary, way to reap the benefits of automated technology sooner. Because the FA employment concept is a long-term vision, we mainly focus on the MM and PU employment concepts.

## Personnel Reductions and Efficiencies Can Be Reached in the MM Employment Concept

Table S.1 compares the personnel reductions and potential efficiencies that can be gained from the MM, PU, and FA concepts.

As can be seen in Table S.1, there is only a difference of 9 percent between the MM and PU employment concepts. This marginal difference is due to the need to carry backup drivers in the PU employment concept. The FA concept provides significant personnel reduction, estimated at 78 percent. These calculations are based on the assumption that the number of flatrack positions in the composite palletized load system (PLS) platoon convoy remains constant despite the personnel reduction, creating a potential efficiency in throughput per soldier.

---

employment concepts rely on the same basic technology requirements; it is the employment of these technologies that differentiates them.

**Table S.1**
**Personnel Reductions and Efficiencies Comparisons for Different**
**Automated Truck Employment Concepts**

| AT Employment Concept | % Decrease in Soldiers (Versus Status Quo) | % Increase in Per-Soldier Throughput |
|---|---|---|
| Status quo | None | None |
| MM | 28% | 38% |
| PU | 37% | 59% |
| FA | 78% | 350% |

NOTES: AT = automated truck. One of the constraints in the MM and PU scenarios is that there must be sufficient drivers with the 88M military occupational specialty (MOS) for all of the PLS trucks.

The technology and DOTmLPF-P analyses, summarized below, suggest that the technology required for the FA and PU employment concepts is not yet mature and that significant force structure alterations will be required to realize this per-soldier throughput increase.

## Technology Assessment Results: PU Still Has Risks; MM Is a Better Option for Now

The research team assessed information concerning AV technology maturity in fields in which a complex driving environment is an absolute. These vehicles include commercial trucks, buses, mining trucks, Army trucks, and passenger cars. The driving environments are summarized in Figure S.1.

The left column in Figure S.1 shows different types of driving environments. These are ordered according to increasing complexity, ranging from test tracks to off-terrain trails. The assessment results suggest that the technology needed to put the PU employment concept into action might reach deployment readiness for highway driving in 2019, at the earliest. For full automation, the FA employment concept will take much longer. The MM employment concept, on the other hand, is ready for Army adaptation and deployment in urban and

**Figure S.1**
**Summary of AV Technology Demonstrations**

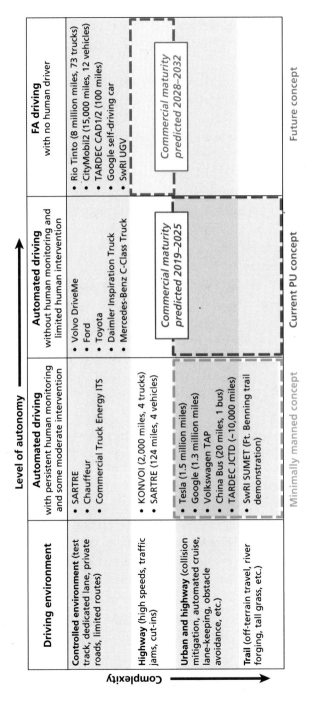

NOTES: CAD = capabilities advancement demonstration; ITS = intelligent transportation system; JCTD = joint capability technology demonstrator; KONVOI = convoy [in German]; SARTRE = Safe Road Trains for the Environment; SUMET = small unit mobility enhancement technology; SwRI = Southwest Research Institute; TAP = temporary auto pilot; TARDEC = Tank Automotive Research, Development and Engineering Center; UGV = unmanned ground vehicle.

highway environments. Moreover, a demonstration by the Southwest Research Institute's Small Unit Mobility Enhancement Technology at Fort Benning suggests that the technology can be developed for trail driving today. This analysis shows that the human operator provides a level of redundancy and robustness required to compensate for the current shortcomings in automated technology.

A technology risk assessment, drawn from a review of test data and input from SMEs, suggests that there are several major technology shortcomings that will likely encumber the development of the PU employment concept. The MM employment concept also contains technical risks, but these risks are more manageable. These risks are summarized in Figure S.2. The risks were classified into seven general categories. Each category ranks the severity of the risk and its probable effect on the development program within the Army. Red risks were assessed to be severe developmental risks due to technology immaturity (technology readiness level [TRL] < 6) or other significant programmatic risks. Orange risks were assessed to be significant developmental risks due to some uncertainty in technology maturity (possible TRL = 6) or other significant programmatic risks. Yellow risks were assessed to have some potential technical and programmatic issues.

## Automated Trucks Will Require Convoy Tasks, Training, and Organization Structure to Adapt

Automated convoys will bring about a dramatic change in the way the Army conducts its combat logistics operations. A qualitative STeS approach was used to guide a structured examination of the likely tactical and force impacts of employing automated trucks in Army convoy operations. This analysis revealed that soldiers who remain in the convoy could have higher cognitive loads as they perform additional tasks. In particular, the span of control might reduce for the convoy commander but increase for the crews of lead vehicles that must manage unmanned following trucks. Lastly, automated trucks will create greater demands for more-senior drivers and fewer demands for entry-level drivers. It is

**Figure S.2**
**Technical Risk Assessment of PU and MM Employment Concepts**

| Critical technical risk | PU | MM |
|---|---|---|
| **Sensors/data fusion:** Inability of sensors/software to correctly interpret and react in complex driving environments | Automated technology ability to correctly perceive and react to hazards remains a major technical risk | Single operator will be available in the cab to monitor and take over when necessary |
| **Sustainment/maintenance:** Inadequate sustainment funds may prevent necessary software upgrades | Inadequate sustainment funds may limit the software and hardware upgrades necessary to improve capabilities | Army can still reduce soldier risk with MM concept if funding is curtailed |
| **Safety/testing:** Impossible to test LF with confidence that it will meet current safety and performance requirements | Millions of miles required for adequate testing, unlikely to occur in development | Single operator allows for accumulation of data fundamental for safety validation |
| **Cyber:** Inadequate cyber mitigation strategies in architecture may increase vulnerabilities and costs to sustain | Jamming of communication and GPS likely will require convoy to stop and reload drivers from other vehicles | MM concept will have single driver in cab to take over if linkage is lost |
| **Communications:** Intermittent or lack of communication between leader and followers will cause instability in followers | Maintaining conformity to prescribed path has technical and safety issues | Follower driver will need to follow leader without benefit of truck commander (TC) as additional observer |
| **Convoy integrity:** Default conformity to following of the leader's path may cause unintended accidents due to degraded driving surface | Cyberattacks may go unnoticed until significant issue occurs | Driver can recognize potential compromise and take back control of vehicle |
| **Human-to-machine interface (HMI):** Ineffective HMI will not allow soldiers to safely and effectively manage automated vehicles | Need to design commander control device (CCD) to help increase awareness and decrease cognitive load of leader TC | HMI technological design and tactical operation with the HMI system is critical for safe and effective single-driver operation |

NOTE: GPS = Global Positioning System; LF = leader-follower.

anticipated that this shift in personnel demand will change the force structure requirements, training, and recruiting for convoy soldiers.

### Convoy-Specific Tasks: Fewer Personnel Mean Higher Expectations

Convoy tasks are likely to undergo a redistribution of functions from humans to machines in both MM and PU concepts. These reallocations are of particular concern because *there will be far fewer soldiers to execute all functions not conducted by the automated truck system or when the automated truck system is not fully functioning.* Many of the affected tasks involve sensing and decisionmaking, which could impose excessive cognitive burden on the remaining soldiers in the convoy. With fewer soldiers to execute all remaining tasks in the automated truck-enabled convoy, technology should be identified to help manage cognitive load limitations of the remaining personnel in the convoy.

### Convoy Organizational Structure: Reporting Structure and Control Will Change

Changes brought about by automated technology will also affect the organizational structure of the convoy for the MM and PU employment concepts—to a greater extent for the PU employment concept. A particular issue that will result from the PU employment concept is related to the direct reporting relationships and their associated span-of-control implications. Currently, almost all trucks have communications capabilities so that truck crews can communicate directly with the convoy commander (CC) or assistant convoy commander (ACC). The addition of unmanned automated truck technology will reduce the number of manned trucks with which the CC and ACC must coordinate. However, the trucks leading the automated unmanned trucks in the convoy will have to provide oversight and management of these unmanned trucks. Although the span of control for the CC and ACC will be reduced, the span of control (and associated cognitive load) for crews in the lead trucks will significantly increase.

### Personnel: Skills Will Change, as Will Training Needs

The proficiencies required for the personnel operating Army convoys with automated technology will also change. Because almost all personnel within the PLS convoy are from the 88M MOS, the introduction of automated trucks will significantly affect the 88M career progression over time. The 88M MOS has four levels that represent progressively increasing levels of knowledge, skills, and abilities (KSAs). Figure S.3 depicts the four MOS levels with the roles generally performed. The roles listed in red text are ones that are likely to have significant impacts from the introduction of automated truck technology.

The vast majority of in-convoy personnel reductions will occur at the 10-level and 20-level 88M positions, with little or no savings among 88M senior noncommissioned officers (NCOs) at the 30 and 40 levels.[3] These changes will reduce the number of soldiers at risk

**Figure S.3**
**88M MOS Pyramid with Key Positions by MOS**

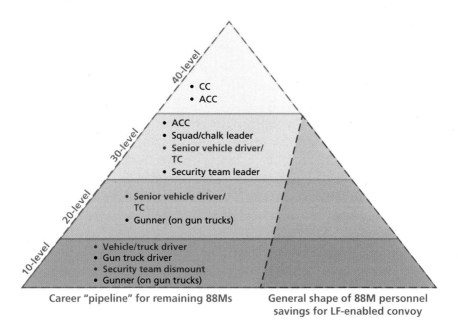

Career "pipeline" for remaining 88Ms

General shape of 88M personnel savings for LF-enabled convoy

---

[3]   These anticipated reductions in personnel needed are only during the actual convoy operation. These soldiers also have responsibilities before and after the convoy operation that will not benefit from the automated technology. The automated technology may actually increase the before and after tasks required to be completed by these soldiers.

but will not directly enable generation of additional convoys and over-all throughput increases. Additional shifts will require reorganization of existing transportation company force structure to increase the 30- and 40-level 88Ms relative to the 10- and 20-level personnel. These demands will eventually change the fundamental structure of the 88M MOS career pyramid and possibly require alternative approaches to training and recruitment for developing senior 88M personnel. The MM and PU employment concepts will have nearly equal effect on these force structure changes.

## Recommendations

Despite the potential pressures and risks associated with being one of the Army's first major automated vehicle programs, the Army should continue pursuing the automated truck technology for convoy opera-tions. This leading, large-scale automation of Army vehicles will be a pivotal effort because robotics are anticipated to be a major force enabler in the future. As such, we provide the following five recom-mendations to help guide this effort.

**Execute the MM employment concept as a necessary bridg-ing strategy to achieve the full PU employment capability.** Current sensor and software technologies do not have the maturity to success-fully manage the complex combat environment to meet the basic tac-tical operating requirements of Army convoys. The human operator, the distinguishing feature of the MM employment concept, provides a level of redundancy and robustness to compensate for current short-comings in automated technology. The driving environment in which automated Army trucks will need to successfully traverse is highly com-plex. Current automated technologies are sufficiently mature if there is a soldier within the vehicle monitoring the driving environment and regaining control of the truck in situations that the automated system is unable to handle. Pursuing the MM employment concept requires nearly all the same technology requirements as outlined in the PU employment concept, allowing the development program to proceed under this current requirement. Further, the MM employment concept allows reducing the number of soldiers in the cab from two to one, which maintains the primary motivation—reducing soldier casualties. The major adjustment, though, is the employment of the automated truck technology with one soldier in the truck. This will be a major cultural adjustment for the Army.

**Ensure that the human factors design is robust in order to mitigate employment risks inherent in the interim MM employment concept.** The major technical drawback to the MM employment concept is the human factors design, or HMI. There are three design and training aspects to the HMI: sensor perception communication, multimodal warning, and external communication. The success of employing automated truck technology under the MM employment concept is highly dependent on effective designs in these three HMI areas.

**Develop clear and practical technical requirements to reduce key development risks.** The risks identified in this study can be mitigated by managing them early in the development process. For example, vehicles are vulnerable to cyberattack. One of the most effective ways to secure a vehicle from such vulnerabilities is to include cybersecurity measures during the initial architecture development.

**Use the MM approach to collect sustained user input for PU development and refinement.** The user will play a vital role in achieving the PU employment concept. Our analysis identified that it will be critical to use the MM approach as a bridging strategy to the PU employment concept. As soldiers conduct convoys using the MM employment concept, valuable feedback and data can be obtained. This information will be critical in improving the automated system to meet soldiers' needs and gain the trust of the force. Without this close partnership, the automated truck will struggle to keep up with the continually changing, dynamic operational environment. Furthermore, during development of the MM technology, user input regarding specific requirements and specifications will be critical.

**Prepare for the inevitable long-term force structure and personnel impacts resulting from automated technology emergence.** Automated technology will pose significant long-term impacts for every aspect of the convoy sociotechnical system. Previous military experience with automation also demonstrates the inevitability of force and, ultimately, workforce changes. Therefore, the pressure to leverage automated trucks to reduce force structure will likely build ahead of system maturation. The Army must be prepared to respond to these pressures with accurate assessments of system readiness and the risks associated with immediate system employment or force structure reductions.

# Acknowledgments

We would like to acknowledge the great support we received across the many organizations of the Army during this research effort. First, we would like to acknowledge our action officer, Keith Jensen, Program Executive Office Combat Support and Combat Service Support, who provided substantial support throughout the study. We would like to thank the U.S. Army Combined Arms Support Command Training and Doctrine Command (TRADOC) Capabilities Manager Transportation—in particular, Don Overton, for all his time and efforts in supporting the activities of our study, especially the support for the two-day workshop we held. Bernie Theisen and Alex Kade from TRADOC provided valuable insights into the study from their numerous research and development efforts in this area. Finally, we would like to thank our reviewers, who provided valuable feedback: Lieutenant General Mitchell Stevenson (ret.) and Maynard Holliday, a senior engineer at the RAND Corporation.

# Abbreviations

| | |
|---|---|
| ACC | assistant convoy commander |
| AV | autonomous vehicle |
| CC | convoy commander |
| CCD | commander control device |
| CSSB | combat sustainment support battalion |
| CTC | composite truck company |
| DoD | U.S. Department of Defense |
| DOTmLPF-P | doctrine, organization, training, materiel, leadership, personnel, facilities, and policies |
| DSB | Defense Science Board |
| EU | European Union |
| FA | fully autonomous |
| GPS | Global Positioning System |
| HMI | human-to-machine interface |
| IED | improvised explosive device |
| KSAs | knowledge, skills, and abilities |
| LF | leader-follower |

| | |
|---|---|
| LIDAR | light detection and ranging |
| METL | mission essential task list |
| MM | minimally manned |
| MOS | military occupational specialty |
| MRAP | mine-resistant ambush-protected |
| MTOE | Modified Table of Organization & Equipment |
| NCO | noncommissioned officer |
| NHTSA | National Highway Transportation Safety Administration |
| PEO CS&CSS | Program Executive Office Combat Support and Combat Service Support |
| PLS | palletized load system |
| PU | partially unmanned |
| SME | subject-matter expert |
| STeS | sociotechnical systems |
| TC | truck commander |
| TRL | technology readiness level |

# CHAPTER ONE

# Introduction

Automated technology is rapidly evolving. Today's prototypes and working models were still in the realm of science fiction only decades ago. Among the most exciting concepts under development are self-driving vehicles.[1] Google's self-driving car project (officially known as Waymo), in which cars use sensors, software, and detailed maps to navigate the road, is probably among the most well known, but a number of major automobile companies, such as BMW, Ford, General Motors, and Tesla, are also committed to developing fully self-driving models. There are many potential benefits of self-driving vehicles. According to one report, dependence on self-driving automobiles may "substantially affect safety, congestion, [and] energy use" for the better in the long term (Anderson et al., 2016, p. xvi).

The U.S. Army, which has thousands of ground vehicles, is also interested in harnessing the potential benefits of self-driving vehicles. Automation, in theory, could save lives by reducing the number of personnel operating in combat zones and increasing convoy efficiencies. Automated convoys are of special interest: Recent combat operations have consistently demonstrated the vulnerability of convoys to attack. These operations are often required to cover extensive distances on unsecured routes. Convoys are particularly vulnerable to attack and ambushes in noncontiguous and noncontinuous combat spaces without generally secure rear areas. Recent operations in Iraq and Afghanistan further illustrated these vulnerabilities with extensive insurgent use of

---

[1] Throughout this report, we use the terms *self-driving vehicles*, *autonomous vehicles*, and all similar constructions interchangeably.

direct fires, coordinated ambushes, and many variations on improvised explosive devices (IEDs) against logistics convoys. Almost all imaginable future scenarios include conventional and hybrid warfare aspects that will pose threats to convoys throughout the entire battlespace.

The Army research and development communities have been testing a concept in which manned and unmanned trucks perform cooperatively in tactical convoy operations. This concept, although promising, needs further examination to realize the full benefits. The technology, although under rapid development in the commercial realm, is not yet ready for full Army deployment. Moreover, it is not fully understood what kinds of operational changes automated trucks will introduce. Thus, it is important that the Army carefully consider the state of the art and potential related changes that automated trucks may introduce in order to manage them with foresight while continuing to advance potential benefits.

Understanding the need for full analysis, the Program Executive Office Combat Support and Combat Service Support (PEO CS&CSS) asked RAND Arroyo Center to assess the risks that automated truck acquisition may experience in development and wider Army operations in the near to mid-term future (one to five years). This research aims to determine the specific risks and risk mitigations for the development of automated trucks. The research team developed two research areas and related questions to address this problem:

- **Technology:** How mature is autonomous vehicle (AV) technology for Army convoy operations? What are the potential risks in deploying this technology for Army convoy operations?
- **Doctrine, organization, training, materiel, leadership, personnel, facilities, and policies (DOTmLPF-P):**[2] What effects will automated convoys have on Army force structure and operation planning and execution?

---

[2] The tactical assessment covers select aspects of DOTmLPF-P but is not a full DOTmLPF-P assessment. For this reason, we use the general term *tactical* instead.

Addressing these questions enabled us to identify opportunities, implications, and risks associated with automated convoys both quantitatively and qualitatively. Perhaps most importantly, the team was also able to develop and assess a short-term concept for the Army to begin implementing almost immediately. This concept and longer-term Army concepts are introduced in the next section.

## Project Scope: Three Concepts for Army Automated Convoys

To date, the Army research and development activities have tested two main automated convoy concepts. The first is what we term the *fully autonomous* (FA) employment concept, which consists of all the cargo vehicles in the convoy being unmanned and driven autonomously. In this concept, all soldiers are removed from the convoy cargo trucks. A remote control station is used to monitor the autonomous driving and manually drive the truck in situations that the automation is unable to manage. Although this unmanned concept is the ideal, the technology is under development, and it may be some time before driverless tactical vehicles can navigate the hazards and obstacles, including road intersections, traffic, pedestrians, and wartime adversaries and threats, in both rural and urban settings.

Understanding this time frame, the Army has research and development activities testing a mid-term solution. This concept is known as the *partially unmanned* (PU) employment concept. In this concept, a palletized load system (PLS) truck is outfitted with an applique kit that allows two soldiers driving a "leader" PLS truck to establish a path for completely unmanned "follower" PLS trucks.

In our initial analysis, we found that this PU employment concept has limiting technical and tactical issues. These findings are presented in Chapter Three, where we describe how current technology is too immature to enact this concept safely, and in Chapter Four, where we describe the tactical risks that would be faced in contingency situations should the PU concept be put into immediate action.

To make a more thorough assessment and to provide the Army with a viable way forward, we developed another concept for analysis: the *minimally manned* (MM) employment concept.[3] This concept is used by many companies developing automated vehicles today and, as we show, should be considered as a bridging concept between today's human-operated convoys and the Army's desired mid-term concept, PU. These two concepts require essentially the same technical requirements for the applique kit. The difference in the two concepts, however, lies in the manning of the follower trucks. In the PU, the follower trucks are completely unmanned. In the MM, there is one soldier in the driver's seat to monitor the automated system and driving environment but no passenger, as in traditional convoy operations. We describe both concepts more fully in Chapter Three.

## Study Methods and Limitations

This study was conducted in four stages and is based on multiple methods. In the first stage, we sought to build context by reviewing current Army convoy operations. Here we focused primarily on the aspects of the operating environment that will guide development and implementation of AV technologies. In the second stage, we assessed the technical benefits and risks associated with the PU employment concept that has been the primary focus of Army research and development. Three sources guided our assessment: expert literature, input from subject-matter experts (SMEs) in various fields related to AV, and test data from Army automated vehicle technology demonstrations and commercial automated test vehicles. It was at this point in the study that we developed and subsequently assessed the MM employment concept with the same rigor as the PU employment concept. In the third stage, we assessed the tactical implications of both concepts through a socio-technical systems (STeS) analysis, which guided a structured exami-

---

[3]   We explicitly use the term *employment concept* instead of *concept* to distinguish that these ideas differ mainly in the ways automated trucks are used. The PU and MM employment concepts rely on the same basic technology requirements; it is the employment of these technologies that differentiates them.

nation of the likely tactical and force impacts of automated trucks. Finally, we developed a set of recommendations based on all study findings. It should be noted that this study does not consider the cost-effectiveness of either concept.

## Document Organization

In Chapter Two, we provide a short background discussion on Army convoys and current Army autonomous systems in general and describe how automated trucks are expected to assist in these missions. In Chapter Three, we describe the two primary concepts chosen for analysis: the Army's PU employment concept and the MM employment concept. In Chapter Four, we present a systematic assessment of current AV technology capabilities and areas of successful application in order to assess the technical maturity and potential risks for automating Army trucks. Chapter Five presents a detailed analysis of salient collective and individual convoy tasks affected by automation to identify and assess expected tactical and force implications of automated trucks. Finally, in Chapter Six, we summarize our findings in all areas and offer several recommendations for PEO CS&CSS and other Army leaders to consider going forward.

# CHAPTER TWO

# An Overview of Army Combat Logistics Patrols and Convoys

The emergence of mature automation and autonomous technologies presents a promising opportunity to significantly reduce risks to ground logistics vehicles and the personnel operating them. However, capitalizing on this opportunity requires a practical approach to maturing, applying, and integrating technology that is consistent with how and where convoys operate. In this chapter, we describe the key aspects of convoy operations and their operating environment. This background information should guide development and implementation of AV technology with the aim to reduce risks to personnel and increase efficiencies. This chapter also presents the key operating environments and tasks of tactical convoys that challenge and constrain the technical capabilities described in Chapter Four.

## Combat Logistics Patrols and Convoy Operations Are Dull, Dirty, Dangerous—and Necessary

In 2007, the U.S. Department of Defense (DoD) recognized the benefits that unmanned systems can offer to the services. These included assistance in mission roles that (1) are long in duration ("dull"); (2) operate in zones that threaten personnel health, such as nuclear sites ("dirty"); and (3) offer extensive risk to human life and political interests ("dangerous") (DoD, 2007, p. 19). Indeed, convoy operations often face all of these conditions and are also fundamental for delivering the materiel required to support combat operations.

Army units conduct convoy operations with a range of vehicles, including large tactical vehicles, medium tactical vehicles, and civilian local-national trucks. The standard convoy consists of PLS trucks, gun trucks, and transportation soldiers. The PLS truck provides an ideal platform for initial implementation of AV technology because it often operates in the environments that pose the greatest risks to soldiers.

## How PLS Convoys Are Composed

Convoys are conducted by Army transportation units that maintain the personnel and capabilities required to execute the convoy operation. One specific unit type is the composite medium truck company (PLS) (also known as a composite PLS company).[1] To provide a standard unit of analysis and comparable outcomes, we use this unit throughout our analysis to assess and compare different technology and employment alternatives.

The composite PLS company consists of PLS cargo trucks, gun trucks for security, and a maintenance and recovery truck. Each of the PLS trucks can pull a trailer as well. Taken together, the composite PLS company trucks can accommodate a variety of modular loads, such as 20-foot shipping containers, water purifiers, boats, mixed cargo supplies, and many other load types. Because of its ability to haul a diverse range of cargos across a wide range of terrain types, the composite PLS company provides an essential capability to support maneuver of tactical units and their associated sustainment units.

---

[1]   Composite truck companies (CTCs) are a new type of transportation unit and are one of the three base units. At its core, the CTC provides motor transport capability to move personnel, containers, flatracks, and heavy equipment under the mission command of the combat sustainment support battalion (CSSB). There are two types of CTCs: light and heavy. The light CTCs are designed to support infantry and Stryker brigades, while the heavy CTC is designed to support armored brigades.

## How Convoys Are Executed

Employment of the composite PLS company to support combat operations requires execution, coordination, and synchronization of numerous actions. Convoy operations require coordinated execution of numerous collective and individual tasks. Introduction of automation will affect the execution of many of these tasks.

We determined that 34 of the total 108 collective tasks in the mission essential task list (METL) for the composite medium truck company (PLS) are likely to be impacted by introduction of the intended technologies. The RAND research team designed and executed user group elicitation to assess the likely implications for execution of these tasks in the automated truck-enabled convoy. Although automated truck capabilities can reduce the number of soldiers at risk in convoys, they also present some new risks due to the need to manage execution of all required tasks and their associated cognitive loads with fewer people than current convoys.

## Where Convoys Operate

Convoy operations are complex under optimal conditions. However, tactical convoy execution can present a particularly vexing set of challenges for Army units. The composite PLS company is designed to operate in a range of environments with varying levels of road infrastructure. There will be different mixtures of physical terrain, built terrain, and local populace. Convoys must cover extensive distances of unprotected routes. Also, convoys often span more than one of the route types described.

In addition to varying physical aspects, the composite PLS company can face a wide range of other conditions across each aspect of the operating environment. Table 2.1 lists the key aspects of the convoy operations environment and provides illustrative examples with varying levels of complexity. Composite PLS personnel and systems must execute the collective tasks described above through many potential variations in the operating environment. This presents challenges for current systems and training, as well as for any future technologies designed to help automate convoy tasks and operations.

**Table 2.1**
**Key Aspects and Examples of the Operating Environment for PLS Convoy**

| Aspect | Description | Examples of Different Levels of Complexity | | |
| --- | --- | --- | --- | --- |
| | | **Low** | **Medium** | **High** |
| Natural/built physical terrain | • Physical character of a piece of ground or area, especially with reference to operational impacts | • Dense forested or jungle conditions | • Some villages; similar to Afghan Ring Road | • Rugged mountain roads<br>• Dense inter-city transit |
| Infrastructure | • The facilities, features, and/or systems that support vehicle operations | • Well-established paved roads<br>• Consistent infrastructure | • Inconsistently paved roads<br>• Inconsistent infrastructure | • Single track/rough roads<br>• Little/no infrastructure |
| Weather/atmosphere | • The state of the atmosphere at a place and time with regard to heat, cold, wind, precipitation, etc. | • Intermittent precipitation/obscuration | • Steady precipitation | • Dense fog, frozen conditions |
| Threat | • An object, actor, or event with ability to generate intentional harm or damage to convoy | • Intermittent small-arms attacks | • Coordinated small-arms, IED, and nonkinetic ambushes | • Pervasive and capable ground/air threat |
| Hazard | • An object, actor, or event with ability to generate unintended harm or damage to convoy | • Routine drivers and pedestrians | • Everyday road distractions | • Chaotic driving/unstructured civilian interaction |
| Electromagnetic | • Interrelation of electric currents or fields and magnetic fields associated with convoy systems | • Generally deconflicted spectrum | • Significant density of signals, service interruptions | • Intentional jamming, spoofing, and service denial |
| Other factors | • Other aspects of the operating environment that can influence convoy operations and/or broader enabling capabilities | • Host-nation policies limiting operation of convoys and AVs | • Local populace hostility of U.S., convoy, and AV presence | • Intentional enemy operation in complex environment to confound technology |

The operational environment for the composite PLS company can often present situations that stress the automated technology. The upper-left picture in Figure 2.1 illustrates an unanticipated obstacle requiring the convoy vehicles to bypass and leave the established road. The upper-right picture depicts several potential obstacles and challenges for an automated truck, including a tight turn, a walking pedestrian, and a roadside object that will constrict suitable paths for the PLS and require effective sensing and obstacle avoidance. The bottom picture depicts a convoy traversing unimproved road surfaces with an oncoming personal vehicle.

**Figure 2.1**
**Operating Challenges for PLS Convoys in Recent Combat Operations**

SOURCES: U.S. Army, 2009; and Creative Commons, 2008.

# CHAPTER THREE

# Three Potential Concepts for Automating Army Convoys

In the Iraq War and operations in Afghanistan, Army convoys sustained heavy casualties as they traversed hundreds of miles on unprotected routes. Automating Army trucks offers the potential to remove soldiers from such dangers. To date, the Army research and development community is testing automated and even FA trucks. In this chapter, we describe the concept that has been the main focus of Army research and development efforts. Also, we develop an alternative employment concept to address current technical immaturity risks. Both employment concepts are the primary focus of the ensuing analyses. We also briefly discuss an FA truck concept, which is not explored in too much depth because it is more of a long-term vision for autonomous trucks in the Army.

## Three Concepts for Army Automation: A Brief Comparison

Before we discuss the PU and MM employment concepts in detail, Table 3.1 compares the personnel requirements and potential efficiencies that can be gained from three concepts for automating Army trucks.[1]

---

[1]  These personnel calculations are focused on convoy operations. This study did not examine the impact that truck automation will have on the broader force structure requirements of composite PLS companies.

13

**Table 3.1**
**Personnel Reductions and Throughput Efficiencies Comparisons for Automated Truck Employment Concepts**

| AT Employment Concept | % Decrease in Soldiers (Versus Status Quo) | % Increase in Per-Soldier Throughput |
|---|---|---|
| Status quo | None | None |
| MM | 28% | 38% |
| PU | 37% | 59% |
| FA | 78% | 350% |

As can be seen in Table 3.1, there is only a difference of 9 percent between the MM and PU employment concepts. This marginal difference is due to the need to have backup drivers ride in the gun trucks—five total instead of the traditional three. The FA concept provides significant personnel reduction, estimated at 78 percent. Furthermore, the number of flatrack positions in the composite PLS platoon convoy remains constant despite the personnel reduction, creating an efficiency in throughput per soldier. The analysis in Chapter Five discusses the force structure changes necessary to realize this per-soldier throughput increase.

## Minimally Manned and Partially Unmanned Convoy Employment Concepts

The MM and PU employment concepts require nearly all the same technical requirements. The tactical employment of these concepts, however, is the real area of divergence between the two.

### Minimally Manned Employment Concept

Figure 3.1 illustrates the MM employment concept.

In the MM employment concept, a leader truck provides the driving path for follower trucks being driven by an automated system. Because most of the driving tasks are being done by the automated system, the truck commander, who normally sits in the passenger seat,

**Figure 3.1**
**Overview of Minimally Manned Concept**

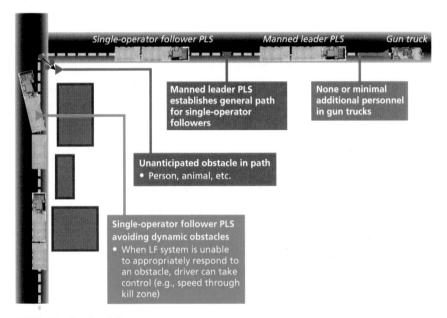

NOTE: LF = leader-follower.

can be removed from the follower trucks. The remaining soldier in the follower truck is best described as an operator instead of a driver. As an operator, the soldier is responsible for monitoring the automated system and driving environment, as well as performing the other tasks normally given to the truck commander. It is anticipated that this single soldier can perform all the necessary functions, potentially with some additional aids, because most of the driving tasks will be done by the automated system. For example, Army tests have shown that the single operator has improved situational awareness while the vehicle is being driven by the automated system (Davis and Schoenherr, 2010, p. 509).

The primary reason that an operator remains in the follower vehicle is to take over driving in situations and conditions that the automated system is unable to handle. A key element to convoy survivability is the ability to quickly pass through dangerous areas and ambushes (Killblane, 2015, p. 21). Yet the sensor and software technology used in today's automated systems still struggles in highly complex situations.

As illustrated in Figure 3.1, dynamic obstacles can pose serious issues for the automated system, resulting in the truck coming to a complete stop for an indefinite amount of time. In these situations, the operator can momentarily regain control of the system and bypass the obstacle. This dependence on an operator after initial development is a common practice in commercial automated vehicle applications and is discussed further in Chapter Four.

### Partially Unmanned Employment Concept

The two main differences between the MM and PU concepts are that, in the PU concept, all the follower vehicles are unmanned and backup drivers ride in the gun truck, as illustrated in Figure 3.2.

The fundamental difference between the PU and MM employment concepts is how shortcomings in the automated system performance are handled. Dynamic obstacles can be difficult for an automated system to properly sense and avoid, causing the follower truck to come to a complete stop, as depicted in Figure 3.2. In this situation, in the PU employment concept, the gun truck can proceed to the disabled unmanned truck and dismount a backup driver. The gun truck normally has three soldiers, providing two extra seats for backup drivers. Once the dismount is completed, the convoy can proceed. Under extreme scenarios in which all of the follower vehicles are unable to proceed, the gun truck can provide backup drivers for two of the follower trucks, and the truck commander in the leader truck becomes the backup driver for the remaining follower vehicle. In this situation, the gun truck will be ferrying between all four of the PLS trucks, including picking up and dropping off the truck commander in the leader vehicle.

### The Technical Requirements for the PU and MM Concepts Are Very Similar

The technical requirements are nearly similar for the PU and MM employment concepts; however, less technical resiliency is required for the MM concept because the soldier operator provides an addition level of robustness against expected technological shortcomings.

**Figure 3.2**
**Overview of the Partially Unmanned Convoy Concept**

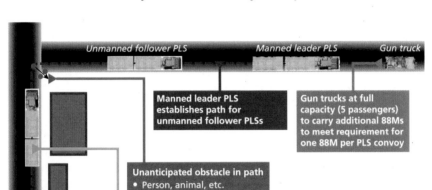

Table 3.2 overviews the general technical requirements for the PU and MM employment concepts.

There are seven general technical requirement areas necessary to develop an automated Army truck capable of following the pathway of a leader vehicle in combat scenarios. Both concepts require technical capabilities to maintain proper gap distances between trucks and path-following for the convoy execution. In both concepts, the commander requires an interface with which to manage the follower trucks in the convoy (e.g., convoy formation, gap distance, situational awareness). Within the follower cab, however, the human-to-machine interface (HMI) is paramount for the MM concept to assist the operator with monitoring the automated system and environment. Even for the PU employment concept, there will be times when a single soldier will be in a follower vehicle requiring similar HMI functions (e.g., after a dismount to a follower truck, the convoy may need to travel some distance

**Table 3.2**
**General Technology Requirements for the PU and MM Employment Concepts**

| Technology Functional Areas | Description |
|---|---|
| Convoy execution | • Number of following vehicles, gap distance, and alignment<br>• Follower vehicles trace path of leader vehicle |
| HMI | • Commander's control device to manage the order of march and situational awareness of following trucks |
| Obstacle avoidance and reassemble | • Capabilities to decelerate or avoid obstacles<br>• Capabilities to align to new leader or serial |
| Sustainment and maintenance | • Ability to sustain operations with minimal interruption due to failures and repairs<br>• Ability to restore automated functionality in timely manner |
| Interoperability | • Ability to incrementally improve capabilities with next-generation sensors and new software updates<br>• LF system does not affect other PLS capabilities |
| Protection systems | • Systems to protect against cyber and electronic warfare threats |
| Safety systems | • Aggregate of safety systems to ensure the safe transport of all types of loads and configurations |

before the backup driver can dismount back into the gun truck).[2] Both concepts will need an acceptable level of performance in detecting and avoiding obstacles for safety and tactical feasibility, more so for the PU employment concept. The technical requirements in the remaining areas of maintenance, interoperability, protection systems, and safety systems are expected to be similar for both concepts.

## The MM Employment Concept as a Bridging Strategy to Achieve the PU Concept

The Army's desire to achieve the level of automation required for the PU employment concept, and eventually the FA employment concept,

---

[2]   There are other HMI technical requirements for the commander's control device that will be similar for both the PU and MM employment concepts.

aligns with the investments and plans of many companies in the private sector. The commercial industry, however, has not reached this level of fully autonomous vehicles. Realizing this goal has proven to require an unwanted but necessary interim step similar to the MM employment concept, as seen in Figure 3.3.

One of the greatest difficulties faced by automated vehicles is their ability to correctly perceive and react to the nearly infinite driving scenarios they may face. The sheer complexity of the potential driving scenarios becomes nearly technically infeasible for the sensor and software technology available. In an effort to develop viable automated vehicles to meet this challenge, commercial companies are limiting the driving environments and/or using human operators as a level of robustness and a necessary component for the continual technology maturation. Depending on a human operator to monitor the driving environment, however, has many unwanted characteristics. Without the human operator, the technology would not be exposed to and learn from the vast array of driving scenarios required to develop a safe and robust system. It is likely that the Army will need to deploy the MM employment concept for an extended period to achieve the desired PU employment concept. In our analysis presented in the following chap-

**Figure 3.3**
**Relationship Between PU and MM Employment Concepts**

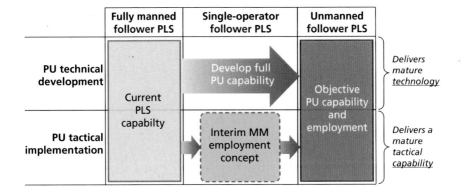

ter, this approach provides a mostly technically mature and tactically practical approach to automating Army trucks.[3]

---

[3]    A competing concept would be to use a remote operating center, where a human would monitor several vehicles and take manual control when the automated system was unable to manage the driving scenario. For this concept to be feasible for the Army, the automation would need to be robust enough to manage driving from divided highway, urban areas, and off-terrain without the need for frequent manual operation. This concept requires a robust communication infrastructure, something the Army cannot assume in a contested environment.

# CHAPTER FOUR

# Technological Assessment: How Close Is the Army to Realizing the Benefits of Automated Convoys?

In this chapter, we systematically assess AV technology capabilities and areas of successful application through 2016, the time the study was completed. This part of the study was conducted to identify the technical maturity and feasibility of automating Army trucks, specifically convoy vehicles, in the near term (i.e., what the Army can start implementing within the next five years, preferably sooner than later). Potential risks were also identified at this stage of the study, as it is important that the AV concept operate successfully and safely in a variety of environments.

## Multiple Sources Informed Technical Review

This part of the study was informed by three main sources of information. Expert literature pertaining to current AV developments and required supporting technologies was reviewed first. Key technological risk areas were also initially identified in this first stage of information-gathering. The second source consisted of in-depth interviews with SMEs in the mining, agriculture, commercial trucking, and academic fields. The topic areas and detailed questions in the interview protocol were derived from the initial insights and outstanding questions derived from the literature review. The protocol we developed was fairly lengthy, but each interview was conducted in such a way that interviewees were asked to respond only to areas in which they had deep subject-matter expertise. Test data from Army AV technology demonstrations and commercial automated test vehicles were our third source. These data sources provided quantitative data to characterize the severity of the identified risks.

## Commercial AV Developments Provide Insight, but Differences Must Be Considered

Over ten years ago, unmanned vehicles traversed a 100-plus mile course in the desert in what has come to be known as the DARPA [Defense Advanced Research Projects Agency] Grand Challenge. Since then, major automakers, truck manufactures, and even Silicon Valley have developed and begun testing automated vehicles. Many luxury cars come with some semiautonomous features, such as lane-keeping and automated cruise control, that relieve the driver of steering and braking on the highway. There is a feel of a "great race" to the autonomous car with all the press attention in this area. As a consequence, many in the Army see the accelerated development of AVs as an opportunity to move soldiers out of harm's way.

Yet it should be remembered that there are significant differences between commercial and military use of AV technologies. For this reason, we begin the technology assessment with a discussion of these differences, as well as similarities. Figure 4.1 frames this discussion, bringing attention to where commercial developments can be harnessed by the Army and where the differences lie.

As the middle box in the diagram suggests, commercial AVs and the projected automated Army trucks will most likely share many of the fundamental building blocks. For example, automated vehicles use a variety of sensors to perceive the driving environment. Common sensors used include radar; optic cameras; and light detection and ranging (LIDAR) sensors, which are laser-based. Currently, many luxury passenger vehicles are equipped with radar and optic sensors to provide semiautomated features, such as lane-keeping and automated cruise control. Industrial vehicles, such as mining equipment, will also

**Figure 4.1**
**Venn Diagram Comparing the Similarities and Differences Between Commercial Automated Vehicles and Army Automated Trucks**

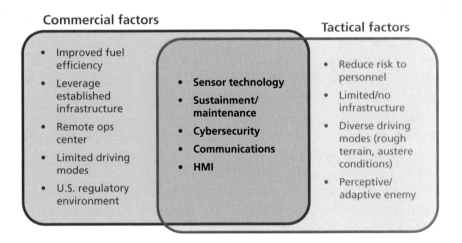

Commercial factors

- Improved fuel efficiency
- Leverage established infrastructure
- Remote ops center
- Limited driving modes
- U.S. regulatory environment

- **Sensor technology**
- **Sustainment/maintenance**
- **Cybersecurity**
- **Communications**
- **HMI**

Tactical factors

- Reduce risk to personnel
- Limited/no infrastructure
- Diverse driving modes (rough terrain, austere conditions)
- Perceptive/adaptive enemy

include the more expensive LIDAR sensor in their sensor suite. As the price of LIDAR sensors continues to drop, it is anticipated that passenger vehicles will begin to be equipped with these sensors. As a result, the Army will have a prime opportunity to purchase these sensors at a competitive price point.

However, the environment in which the automated Army truck will need to operate will be much more complex than AVs operating in the commercial sector. This will require the Army to develop capability in areas beyond the commercial industry. The combat environment has many more features that the automated truck system must account for. These include challenges related to topography (desert, jungle, forest), weather (arid, snow/ice, rain), infrastructure (road surfaces, lane widths, proximity of buildings), obstacles (pedestrian attire/behavior, bicycle density, traffic flow/behavior, types of animals), and, of course, adversary intent and capabilities (sensor spoofing, cyberattack, kinetic attack). Despite the opportunity to use the same sensors as the commercial industry, the Army will most likely need to develop more-advanced software to perceive and react appropriately to the many additional complexities in the combat environment.

## AV Technology: Highly Dependent on Human Operators for Now

We conducted a survey of demonstrated AV technologies across several vehicle types: passenger cars, commercial trucks, buses, mining trucks, and Army trucks. This survey provides a high-level assessment of the environments and conditions in which current AV technology is mature. Although these applications do not align exactly to the Army's requirements for automation, they provide a general indication of technology maturity. Analysis of this survey data revealed that current applications of AV technology in complex driving environments require human operators within the vehicle to monitor the automated system and driving environment. The human operator provides a level of redundancy and robustness to compensate for automated technology shortcomings. Many companies are developing automated vehicles intended to operate without a human sitting in the driver's seat actively monitoring the environment. However, these applications are still limited to the test track and other highly controlled environments.

### The Minimally Manned Concept May Be Feasible

These observations indicate that there are potentially major technology shortcomings for the PU employment concept. However, the technology for the MM concept may be sufficiently mature for development. Survey results are presented in Figure 4.2.

This survey of commercial and Army AV technology provides a high-level assessment of the environments and conditions in which this technology has successfully been employed. The rows describe driving environments ordered in increasing complexity, from highly controlled environments, such as test tracks, to off-terrain applications. The columns describe increasing levels of autonomy that align to the MM, PU, and FA employment concepts.

The first column in Figure 4.2 aligns to the MM employment concept. This column captures applications in which the automated system is driving the vehicle but requires monitoring from a human sitting in the driver's seat. This monitoring is required because the technology is not robust enough to confidently and safely handle all the

**Figure 4.2**
**Survey of Commercial Truck, Commercial Passenger Vehicle, and Army Demonstrations of Automated Vehicle Technology**

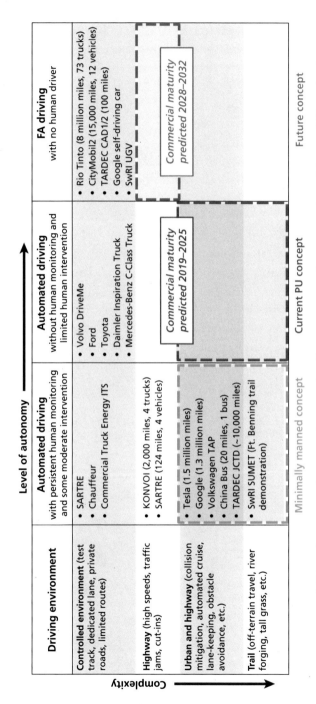

**Level of autonomy** →

| Driving environment | Automated driving with persistent human monitoring and some moderate intervention | Automated driving without human monitoring and limited human intervention | FA driving with no human driver |
|---|---|---|---|
| Controlled environment (test track, dedicated lane, private roads, limited routes) | • SARTRE<br>• Chauffeur<br>• Commercial Truck Energy ITS | • Volvo DriveMe<br>• Ford<br>• Toyota<br>• Daimler Inspiration Truck<br>• Mercedes-Benz C-Class Truck | • Rio Tinto (8 million miles, 73 trucks)<br>• CityMobil2 (15,000 miles, 12 vehicles)<br>• TARDEC CAD1/2 (100 miles)<br>• Google self-driving car<br>• SwRI UGV |
| Highway (high speeds, traffic jams, cut-ins) | • KONVOI (2,000 miles, 4 trucks)<br>• SARTRE (124 miles, 4 vehicles) | | |
| Urban and highway (collision mitigation, automated cruise, lane-keeping, obstacle avoidance, etc.) | • Tesla (1.5 million miles)<br>• Google (1.3 million miles)<br>• Volkswagen TAP<br>• China Bus (20 miles, 1 bus)<br>• TARDEC JCTD (~10,000 miles) | | |
| Trail (off-terrain travel, river forging, tall grass, etc.) | • SwRI SUMET (Ft. Benning trail demonstration) | | |

*Commercial maturity predicted 2019–2025*

*Commercial maturity predicted 2028–2032*

Complexity →

Minimally manned concept     Current PU concept     Future concept

NOTES: CAD = capabilities advancement demonstration; ITS = intelligent transportation system; JCTD = joint capability technology demonstrator; KONVOI = convoy [in German]; SARTRE = Safe Road Trains for the Environment; SUMET = small unit mobility enhancement technology; SwRI = Southwest Research Institute; TAP = temporary auto pilot; TARDEC = Tank Automotive Research, Development and Engineering Center; UGV = unmanned ground vehicle.

potential driving scenarios it may face in the intended driving environments. Europe has been the leader in developing automated driving technology for commercial trucks. The 2016 European Union (EU) truck platooning challenge represents an accumulation of its progress thus far. In this challenge, six different truck platoons drove approximately 3,700 miles across Europe (Figure 4.3).

The European truck platoon is an application with many similarities to the MM concept. However, as can be seen in Figure 4.3, the driving environment in which these European truck platoons are traveling is much more benign than what an Army convoy will likely experience. In the EU truck platooning challenge, each truck platoon consisted of a leader truck providing speed and pathway to following trucks. The drivers in the following trucks were relieved of many of their driving tasks. However, the automated technology was not mature enough to handle all the situations that occurred across the route, requiring the drivers to actively monitor the automated system and driving environment. Drivers reported that they needed to take back control from the automated system in dense traffic, roadway junctions, construction zones with narrow lanes, heavy rain, and some urban situations (European Union, 2016).

**Figure 4.3**
**Map of Routes Traveled by EU Truck Platoon Challenge and Photo of One of the Platoons**

SOURCE: European Union, 2016.

The second column in Figure 4.2 contains applications in which the automated system is robust enough to eliminate the need for active human monitoring, but drivers are still required to drive in prescribed situations (e.g., transitions from highway to urban streets). This column aligns to the PU employment concept because the unmanned following vehicles are intended to operate without soldier drivers, but the convoy carries sufficient backup drivers in case the convoy encounters a situation in which the automated system is not intended to drive (e.g., lingering dynamic obstacles). The trucking industry envisions the truck driver becoming a logistics manager under this level of automation. As a logistics manager, the truck driver will conduct other tasks while the automated system is driving the highway route (Mercedes-Benz, undated). The applications seen in this level of automated driving are still limited to controlled environments, such as test tracks, or limited routes for short demonstration purposes. As can be seen in Figure 4.2, there are no known applications seen in more-complex environments, such as public highways and urban roadways.[1] Commercial developers have made public announcements that their applications will be ready for open highway operation between 2019 and 2025 (Driverless Car Market Watch, undated). The Army is targeting 2019 to start development of automated trucks (Lee, 2016), the same time at which this higher level of automation is anticipated for open highways. However, the Army will require its automated system to operate in much more complex environments.

The third column in Figure 4.2 contains FA applications with no need for a human to be in the truck for monitoring or as a backup driver. However, these concepts depend on remote operators monitoring the autonomous driving and manually controlling the vehicles when the automation is unable to manage. This column aligns to the FA employment concept, in which all the Army cargo trucks are unmanned and the convoy does not carry backup drivers, though there is a remote operat-

---

[1]  Many companies are striving for this level of automated driving and will report their efforts in this area, giving the impression that this level of automated driving is more mature than it actually is. Care should be taken to determine the robustness of the automated system in managing all situations during the tests. Reviewing test results will show that "ghost" drivers are closely monitoring the automated driving and taking back control in many instances.

ing center monitoring the driving and manually controlling the vehicles when necessary. The most impressive example in this category is the Rio Tinto autonomous mine trucks. These autonomous trucks have traveled over 2.4 million miles since 2012 (Rio Tinto Operations Centre, 2014). The stark difference between the Rio Tinto mining operation and the Army combat environment is the high degree of control that is available in a mining operation. Because mining operations are on private property, all of the roads and traffic can be managed and controlled in such a way that the AVs can successfully operate. A combat environment does not allow for this luxury and must deal with many unknowns with local population, herds, and an active enemy.

## The Minimally Manned Concept Reduces Significant Technology Risks

In this section, we discuss the main technical risks that are likely to face the development of the PU or MM employment concepts. These risks and their severity were derived from interviews with SMEs and literature in the commercial truck, commercial automobile, mining, agriculture, academic, and military areas, as well as our review of commercial and Army test results of automated vehicles and trucks. The identified risks were classified into seven general categories. Each category ranks the severity of the risk and its probable effect on the development program within the Army. Red risks were assessed to be severe developmental risks due to technology immaturity (technology readiness level [TRL] < 6)[2] or other significant programmatic risks. Orange risks were assessed to be significant developmental risks due to some uncertainty in technology maturity (possible TRL = 6) and other significant programmatic risks. Yellow risks were assessed to have some potential technical and programmatic issues. Figure 4.4 contains the results of our risk assessment, followed by detailed discussion of each risk.

---

[2]  DoD uses TRLs to determine the maturity of a technology for military development. A TRL of 6 is required to start a program to develop the technology. At the TRL 6 level, the critical components of the technology have been successfully demonstrated in relevant environments.

**Figure 4.4**
**Technical Risk Assessment of PU and MM Employment Concepts**

| Critical technical risk | PU | MM |
|---|---|---|
| **Sensors/data fusion:** Inability of sensors/software to correctly interpret and react in complex driving environments | Automated technology ability to correctly perceive and react to hazards remains a major technical risk | Single operator will be available in the cab to monitor and take over when necessary |
| **Sustainment/maintenance:** Inadequate sustainment funds may prevent necessary software upgrades | Inadequate sustainment funds may limit the software and hardware upgrades necessary to improve capabilities | Army can still reduce soldier risk with MM concept if funding is curtailed |
| **Safety/testing:** Impossible to test LF with confidence that it will meet current safety and performance requirements | Millions of miles required for adequate testing, unlikely to occur in development | Single operator allows for accumulation of data fundamental for safety validation |
| **Cyber:** Inadequate cyber mitigation strategies in architecture may increase vulnerabilities and costs to sustain | Jamming of communication and GPS likely will require convoy to stop and reload drivers from other vehicles | MM concept will have single driver in cab to take over if linkage is lost |
| **Communications:** Intermittent or lack of communication between leader and followers will cause instability in followers | Maintaining conformity to prescribed path has technical and safety issues | Follower driver will need to follow leader without benefit of TC as additional observer |
| **Convoy integrity:** Default conformity to following of the leader's path may cause unintended accidents due to degraded driving surface | Cyberattacks may go unnoticed until significant issue occurs | Driver can recognize potential compromise and take back control of vehicle |
| **HMI:** Ineffective HMI will not allow soldiers to safely and effectively manage automated vehicles | Need to design commander control device (CCD) to help increase awareness and decrease cognitive load of leader TC | HMI technological design and tactical operation with the HMI system is critical for safe and effective single-driver operation |

NOTE: GPS = Global Positioning System; TC = truck commander.

## Comparison of Technology Risks for the PU and MM Concepts

**Sensor and data fusion:** The most prevalent technical risk that will likely face the Army in the development of automated trucks is the nuisance shortcomings in the sensors and data fusion. This risk will limit the ability of the automated system to properly recognize and react to obstacles that will come in a multitude of diverse forms across numerous driving situations. This technical shortcoming will be problematic for the PU employment concept. Follower vehicles will either stop or crash in situations when the automated system is not capable of managing the situation. The technology should be sufficiently mature under the MM employment concept because the soldier will be part of the system resiliency; however, this introduces significant human-factor issues that will be discussed below.

The field of computer vision still faces challenges with how images of any object can change with their pose/orientation and lighting effects. Although LIDAR presents long-range surface detection with reflectivity measures, it has great limitations with presenting any other features that can aid with object identification, activity classification, or predicted intent. The higher the level of perception, the better an autonomous follower vehicle can independently predict and react to a dynamic threat or obstacle that may be encountered on its directed path. The feature detection and perception of static obstacles may be even more complex because dynamic obstacles may be dismissed from feature detection processing if they are found to be removed from the zone of a directed path ahead. SMEs we interviewed identified sensor perception as a major technical issue. Army tests of automated trucks have revealed similar issues seen in commercial testing of automated vehicles (Heim, 2015, p. 30). Issues related to perceiving the environment have been the prevalent reason for operator takeovers of the Google automated car system (37 percent of all takeover events) (Google Auto LLC, 2015, p. 10).

**Safety and testing:** Testing of automated vehicles poses programmatic and technical risks for the Army. The sheer complexity of driving environments and scenarios in which the automated system will need to operate will tax even the best-resourced testing program. Resources for demonstrating and testing technologies are continually being constrained, resulting in systems being deemed nonoperationally suitable later in the program (Hunter et al., 2016). Furthermore, the Army will need to ensure that it has sufficient technical capabilities and test facilities to test the software complexities of automated systems. These issues pose significant risks for the PU employment concept if there are limited testing resources and capabilities. Under this scenario, there will most likely be a limit to the likelihood that the automated system will be robust enough against the uncertainties in the combat environment. The MM employment concept provides a layer of robustness because the soldier operator will be the fallback in situations when the automated system is unable to manage the situation. However, this approach will have human-factor issues that must be properly managed.

Both the commercial and DoD communities face similar technical challenges for testing automated vehicles. A 2012 Defense Science Board (DSB) study concluded that DoD requires new technical capability for testing the complex software systems inherent in automated systems beyond what is normally required for most combat systems (DSB, 2012, p. 9). In addition, this study concluded that DoD will need to improve its operational test ranges so that it can better evaluate autonomous systems (DSB, 2012, p. 12). The commercial sector faces a similar issue. It is estimated that automated vehicles would need to be driven hundreds of millions of miles to clearly demonstrate their safety and effectiveness. To overcome this statistical complication, the commercial sector will need to innovate other methods, which may be of use to the Army, to ensure the successful and safe integration of automated vehicles. Such methods include modeling and simulation, accelerated testing, and scenario testing (Kalra and Paddock, 2016).[3]

**Sustainment and maintenance:** The installation of an applique kit will increase the complexity of the vehicle, requiring sufficient sustainment funding to manage software updates and mechanics with additional skill sets and competency. Limited resources for software

---

[3]    The commercial sector is extensively using public roadways to conduct its testing of automated vehicles and using simulation to augment its roadway tests. The Army is actively pursuing these areas also, which presents another area of opportunity for the Army to learn from and potentially leverage developments in the commercial sector.

upgrades become problematic for the PU employment concept because the automated system may lag in its ability to meet the operational needs of the soldier. For the MM concept, the automated system may still be functional with the aid of the soldier within the cab.

Sufficient sustainment funding will be required not only to maintain the automated system but also to support the additional software enhancements after initial fielding. The additional sensors, electronics, and software required for the automated system will naturally increase the sustainment costs. For passenger vehicles with automated technology, it is estimated that maintenance costs will increase by a few hundred dollars annually (Litman, 2013). A significant portion of this estimated cost is for the software updates and information required for the continued safe and efficient operation of the vehicle, mostly likely paid by the developer. These software enhancements are necessary because of the inability to fully discover, incorporate, and test all the logic necessary during development.

Inadequate maintainer training and competency to diagnose and repair the automated system may result in low readiness or availability. The automated system consists mostly of electronic and software components, which are much more difficult to diagnose than their mechanical counterparts. The automated system will also integrate with many other systems in the vehicle, further complicating diagnosis. In addition, over the life of the automated truck, there can be accelerated wear to the brakes due to hard braking events if placed in an area of operation with other road users encroaching on the leader vehicle's path.[4]

**Cyber:** The automated system will need to protect itself against cyber threats. One method to protect against the cyber threat is to design the system so that it is isolated or air-gapped. However, this approach creates potential programmatic risks of schedule delays and cost overruns if the automated system is unable to operate when off the network. Furthermore, there are technical difficulties in protecting the vehicle's electronic systems, as seen by hacker demonstrations of how to launch a cyberattack against a vehicle (Greenberg, 2015). There

---

[4]   Interview with AV engineer, 2016.

are marginal differences in the potential impacts these risks may pose between the PU and MM employment concepts. The soldier operator in the MM concept may be able to recognize that the vehicle is compromised and regain control or bring it to an emergency stop.

The separation of the automated system from the network is technically possible, but it is improbable, given that the automated system will likely reside on the vehicle's communications bus to control the throttle, brakes, and other critical systems. If the automated system is connected to the bus, then any other system on the bus that is connected to the network implicitly makes the automated system connected to the network. Other systems that may be useful to the automated system, such as vehicle metadata collection agents and diagnostic sensors, are useful in identifying anomalous conditions that could imply cyberattacks. However, it may be challenging to collect and act on this data in real time on board the vehicle. It is likely that for this information to be useful, it will need to be shunted into the Army cloud for analysis and alerting of proper stakeholders. If these issues are realized late in the program, schedule delays and cost increases may result from the need to incorporate new information assurance requirements into the automated system.

Protecting the automated system against cyberattack will present technical challenges that can best be mitigated early in development. For example, GPS provides an "open hole" in security, but it is needed by automated and autonomous vehicles for position, navigation, and timing.[5] The technology used to protect the automated system will need to ensure that the operating system running any device within a vehicle is trusted and has not been tampered with. Solutions being developed in the commercial automobile industry require designing a secure architecture early in the development. Examples include "secure boot" technology to ensure that only authorized software is running (Vyas, 2016), "trust-anchor" technology (Reinhold, 2016), Symantec's

---

[5]    Unmanned aerial vehicles are able to use network timing protocols in GPS-denied environments and optical flow for location. For ground systems, however, precision of a few inches is required (e.g., it would not be acceptable for an unmanned PLS truck to run over a curb with people on it while making a tight turn in an urban environment). There is current technology development that may close this gap for ground systems soon.

"code signing and secure boot" (Witten, 2016), and Delphi's "authenticated boot" (Krzeszewski, 2016).

**Communications:** Each automated following truck requires continuous and accurate navigation instructions from the leader vehicle to maintain a precise following path and gap distance. Interruption in this communication can be tolerated as long as the error propagation that results does not cause an incident (e.g., drift does not cause the follower truck to go off the roadway). Intermittent communication may be problematic for the PU employment concept because communication jamming may be common in a combat zone. This issue is much less of a risk for the MM employment concept as long as the soldier operator receives a notice of the jamming and retakes control of the vehicle.

Each following truck in the convoy will need timely and accurate navigation information. Information gained by the first truck needs to be available to the last follower truck. Automated vehicle developers in the private sector have reported issues with intermittent communications.[6] When communication between the leader truck and the follower vehicles is compromised, disturbances and errors are amplified, and following becomes unstable.[7] In addition, communications between trucks are needed to handle nonmilitary cut-ins to the convoy. Most important, communications are critical in conflict situations. Breakdowns in communications could leave trucks idle in the kill zone, a situation to be avoided. In past conflicts, lack of communications has resulted in trucks being left in the kill zone for extended periods (Killblane, 2014, p. 167). Adding automated trucks into the mix may further compound this situation.

The Army's use of radios and radio frequencies that are commonly available is convenient but presents the risk of enemy interference. Frequency agility in radios is very important to defeat enemy interference. For AVs, what information must be transmitted and with what frequency, or finding the optimal frequency, makes radio communication complex. Specifically, "with multiple vehicles, the architecture

---

[6]  Interview with AV engineer, 2016.

[7]  Interview with AV engineer, 2016.

of communications is not solved for ubiquitous, simultaneous information delivery."[8] Fully autonomous vehicles will require more than line-of-sight communications and navigation. If an automated truck is following a leader truck in a crowded urban area, and the leader truck makes a turn and is obscured by buildings, the following truck might come to a complete halt if it has lost its line-of-sight communications link. This could create problems for the following trucks in the convoy, causing them to halt also, if they are operating autonomously.

**Convoy integrity:** The follower truck in the Army convoy is required to follow the path provided by the leader truck with a certain amount of longitudinal and lateral precision. As we discussed above, communications can limit the ability of a follower truck to maintain this path precision. Even with perfect communications, other issues arise with sensor accuracy and road surface conditions that may skew the path of the follower vehicle. For the PU employment concept, this risk may be problematic while traversing narrow lanes or when close to static obstacles. The automated follower truck may strike an obstacle, slip off the road, or come to a stop until a soldier is brought to the truck. Under the MM employment concept, the soldier operator will need to be actively monitoring the automated system and intercede if the truck is dangerously drifting off of the intended path. Under this scenario, the convoy will continue uninterrupted.

The leader vehicle path selection and speed of travel may inadvertently cause a rollover risk or other incident during a loss of traction. In automated truck applications in the mining and agriculture sectors, terrain detection for traction and stability control has caused problems.[9] This risk is faced when traveling over deformable surface terrain (i.e., sand, mud, or gravel). Army tests to date on automated trucks have not tested or done analysis on enhanced roll and yaw stability control, bringing further concern that this technology may not be mature for Army application (Heim, 2015).

Another risk that most likely will be a concern is the true path-following performance of the automated trucks. Although implemen-

---

[8]    Interview with Army engineer, 2016.

[9]    Interview with AV engineer, 2016.

tations and fielded operations in the mining and agricultural sectors have yielded efficiencies in performance, there is still a source of variability due to how people perceive the abilities of the automated system. A vehicle that follows waypoints will track the points directly off of a bridge if directed on such a path. Conversely, when the automated vehicle system traverses a path with other road users, it may not have the features needed to track obstacles in a way that predicts a nearby vehicle's or pedestrian's likelihood and intent to cut through the convoy serial path. The convoy performance may degrade within a mission if the convoy commander does not maintain an awareness of the limits to the follower vehicle's automation.

**HMI:** There are three areas of risks concerning human factors: the interface in the leader truck used to manage the following trucks, the interaction between the automated follower trucks and surrounding traffic and people, and the interface within the follower truck responsible for keeping the soldier operator engaged in monitoring the driving environment. For the MM employment concept, all three of these areas are a concern and present risks, especially the requirement for an operator to be in the driver's seat of the follower truck responsible for monitoring and responding to the driving environment. Out of all the major risk areas, this is the one risk area where the PU employment concept has less risk than the MM concept.

The design and employment strategy of the HMI in the follower truck will present significant technical and tactical challenges. Reaction times of the operator in the follower truck will need to be equal to or better than reaction times observed without the automation. Naturally, the operator in the following vehicle can become easily distracted when relieved from many of the driving tasks that require constant attention. Furthermore, the single operator in the follower truck under the MM employment concept will be responsible for the additional tasks normally managed by the truck commander (the soldier who rides in the passenger seat). Such tasks include radio communication and maintaining situational awareness (i.e., looking out for hazards and threats, such as IEDs). However, the need to maintain vigilance for hazards and threats provides a natural mechanism to keep the soldier operator engaged in monitoring the outside environment.

Furthermore, Army tests have shown that operators who are aided by automated driving are able to identify threats with shorter reaction times (10-percent improvement) and increased accuracy (6-percent improvement) than drivers who are not aided by automation (Davis and Schoenherr, 2010, p. 501).

Human-factors experimentation with automated systems shows that the HMI design must have effective multimodal alarms, ideal placement of equipment required to support secondary tasks, well-trained operators, and operators with high working memory capacity. Table 4.1 compares reaction times with and without automated systems in situations where different alert systems, operator training, and planned distractions are tested.

In Table 4.1, there is one automated driving scenario that has similar reaction times with normal driving. The first row provides the baseline of comparison for normal driver reaction times, which range from 1.2 to 1.5 seconds from hazard presentation to human physical response (e.g., ball in road to foot on brake) (NHTSA, 2002, p. 10). The second row presents experimental results on the reaction time

Table 4.1
Reaction Time Comparison of HMI Configurations

| Driving Mode | Reaction Time in Response to Hazard or Alarm | Distance Covered (at 55 mph) |
|---|---|---|
| Manual driving | 1.2 to 1.5 sec | 30–37 meters |
| Monitoring automated driving while engaged in nondriving tasks (e.g., sending a text) with effective alarm | 1.3 sec | 32 meters |
| Not monitoring automated driving while engaged in nondriving tasks (e.g., working on tablet) with effective alarm | 2.3 sec | 49–61 meters |
| Monitoring automated driving while engaged in nondriving tasks (e.g., sending a text) with ineffective alarm | 4.8 sec | 118 meters |
| Monitoring automated driving while engaged in nondriving tasks (e.g., sending a text) with no alarm | 5.7 sec (for the 46% of operators who responded) | 140 meters |

when the operator is trained to monitor the environment but has some secondary tasks to complete (e.g., send a text). To aid the operator, an effective alarm system is built into the automated system to warn the operator of a potential hazard. As an example, the Google car has a conservative alarm system that requests operator takeover in situations in which the system has uncertainty in the sensor readings or its perception of the environment (which accounted for nearly 80 percent of total disengagements) (Google Auto LLC, 2015). Experimental results in this scenario show that the reaction time is similar to reaction times in normal driving conditions (Blanco et al., 2015, p. 33). However, the experimental results revealed instances of grave concern if there are ineffective alarms (a reaction time of 4.8 seconds),[10] when the operator is not trained to monitor the driving environment (a reaction time up to 2.3 seconds), and when the operator is engaged in nondriving tasks with no alarm (a reaction time of 5.7 seconds for the instances when there was a response, which only happened 46 percent of the time) (NHTSA, 2002, pp. 6, 7, 17). Even with effective alarming, a lingering risk is habituation (decrease in response to a stimulus after repeated presentations) because many alarms would be benign due to the nuisances in sensor performance.

It is improbable that the automated system will provide an alarm in all situations; however, research has shown that proper design of equipment used to support secondary tasks can maintain reaction times within tolerable levels, and operators with a high working memory capacity can reduce reaction times. The Mobius system is a prototype technology that places displays and interfaces in such a way that it keeps the operator's hands and gaze angle in ideal positions (i.e., hands on the steering wheel and gaze out in front of the vehicle). Reaction times while engaged in secondary tasks with the Mobius system increased only 0.1 second from a baseline scenario of monitoring the driving environment without the distraction of secondary tasks. This

---

[10] Multimodal alarms have been shown to greatly reduce reaction time of operators of a vehicle being controlled by an automated system. These alarms use some combination of light, sound, or haptic indicator (e.g., vibration) to alert the operator. When the alarm is limited to one mode, reaction times greatly increase.

baseline was 1.8 seconds, a slightly longer reaction time than seen in normal driving (i.e., 1.3 to 1.5) (Diederichs et al., 2015, p. 2). Individuals with higher working memory capacity are able to manage more-complex, simultaneous cognitive tasks. Experimentation has shown that individuals with higher working memory capacity are able to respond faster to the onset of a hazard (McCarty et al., 2016, p. 1745).

## Main Trade-Offs Between PU and MM Risks

In this chapter, we assessed the technical risks associated with the PU and MM concepts. The main trade-off between these concepts deals with how the follower trucks will be recovered in situations when the autonomous system is unable to navigate. Under the PU employment concept, backup drivers will need to be ferried to the disabled follower vehicles. In the MM employment concept, the backup driver is already within the vehicle.

An additional trade-off deals with the frequency of incidents (unintended halts and accidents) and the burdens involved with recovering from these incidents. Under the PU employment concept, it is anticipated that follower vehicles will halt fairly often or even crash, especially during the initial years of use. Recovery resources and time will be needed to manage these occurrences. The problems associated with unintended halt issues could likely be resolved with extensive testing; however, a limited testing budget may preclude or delay resolution because testing could require millions of miles of unmanned travel. We believe that the MM approach provides a necessary bridging strategy. With a monitor/operator in the vehicle, unintended halts are reduced in frequency and, when they do occur, can be resolved much more quickly. Moreover, the MM concept provides a means of operating the automated system for the miles required to identify and correct the sensory perception issues that are the cause of the unintended halts, thus allowing for the PU concept to be phased in as an evolutionary development.

# CHAPTER FIVE

# DOTmLPF-P Assessment: What Changes Will Automated Convoys Bring to Army Operations?

Previous chapters illustrate the motivation for and potential benefit of automated trucks to reduce the risks and improve other aspects of tactical logistics convoys. Although the various potential approaches to implementing automated trucks each present a discrete set of potential benefits, all present major implications for Army forces conducting operations. These likely impacts include tactical execution of collective and individual tasks, as well as broader impacts for Army force structures and the personnel who compose them. This chapter examines key implications in the tactical and force spheres that the Army will need to consider and address as part of any development plan.

## Mixed Methods Were Used to Assess DOTmLPF-P[1] Implications

Because of the diversity, complexity, and interrelated nature of the likely impact of automating Army trucks, we applied a multidomain analytic method to examine the various aspects. The qualitative STeS approach guided a structured examination of the likely tactical and force impacts of automated trucks. The section below provides a brief description of the methods used to assess the range of likely implications.

---

[1] We do not do a complete DOTmLPF-P assessment here; instead, we examine key components that are likely to have a significant impact from automation.

## Sociotechnical Systems Approach for Assessing Broader Impacts of Automated Truck Implementation

The STeS construct is particularly useful because it specifically focuses on how machines and humans distribute and structure tasks in a single system. The STeS construct consists of a set of interdependent systems and capabilities that, in addition to technical systems, includes operational processes and the people who use and interact with the technical system. STeS are defined by the following characteristics (Sommerville, 2011):

- The system consists of a purposeful collection of interrelated components working together to achieve a common objective.
- Properties of the system as a whole depend on the system components and their relationships.
- The system may include software, mechanical, electrical, and electronic hardware and can be operated by people.
- System components are dependent on other system components.
- The properties and behavior of system components are inextricably intermingled.
- The system includes technical systems but also operational processes and people who use and interact with the technical system.
- The system does not always produce the same output when presented with the same input because the system's behavior is partially dependent on human operators.

Given the defining STeS characteristics, an STeS approach to organizational development seeks to optimize human resources and technical systems based on their comparative benefits and limitations (Weisbord, 1991). The STeS approach provides "menus of choice" within a simple, comprehensive, and flexible vocabulary to characterize system elements and their connections that change. For these reasons, the STeS construct provides a particularly useful structure with which to guide systematic consideration of the automated truck technology's impact on the tactical PLS convoy as a STeS and the secondary impacts to the forces that compose the convoys. Figure 5.1 illustrates the STeS construct and how its aspects are manifested in the PLS convoy. The

**Figure 5.1**
**The Traditional Sociotechnical Systems Approach as Applied to Examination of Autonomous Vehicle Convoy Operations**

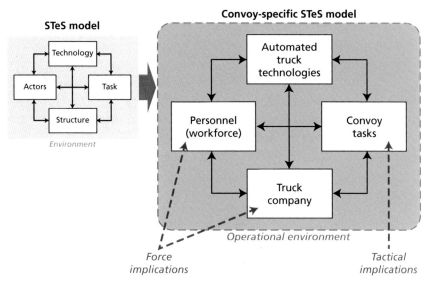

SOURCE: Adapted from Lyytinen and Newman, 2008.

figure also indicates which STeS aspects are most closely related with tactical and broader force impacts.

## Sociotechnical Implications of Autonomous Trucks

Because of the dramatic change in capability that automated trucks will represent, there are significant implications for each aspect of the STeS that the Army will need to consider and address in tandem with system implications. This section describes the analysis and identified key impacts for each STeS aspect.

### Convoy Tasks

The introduction of automated trucks is primarily intended to reduce the number of personnel that Army units must put at risk to execute logistics convoys. The technology is planned to perform many of the driving tasks that must be performed now by the soldier. This will constitute a redistribution of functions from humans to machines. However, the impact of this function redistribution will not be even across tasks and may even generate new functions. Assessing implications of automated trucks to tasks requires an examination of all collective and individual tasks. To identify key impacts and illustrate the approach required for more-detailed analysis, we applied criteria to prioritize collective tasks whose execution is most likely to be significantly impacted by automated truck technology. The PLS truck company has 108 collective tasks on the unit METL. However, only 34 of these tasks will likely need to be augmented by the autonomous technology.

To identify likely impacts of automated trucks across these functional areas, RAND Arroyo Center convened a workshop with SMEs and current practitioners. The group focused specifically on a selection of ten tasks identified to be significantly affected by automated truck technologies across four functional areas: organizational control, convoy operation, convoy security, and maintenance. These tasks formed the basis of our subsequent analysis.

We found that many of the tasks significantly impacted by the automated truck capability are essential and will remain after introduction of automated trucks. However, significant reallocation of functions between the soldier and the system will be required. These reallocations are of particular concern because there will be fewer soldiers to execute all functions not conducted by the automated truck system or when the automated truck system is not fully functioning. Many of the affected tasks involve sensing and decisionmaking, which impose extra cognitive burden on the soldier. With fewer soldiers to execute all remaining tasks in the automated truck–enabled convoy, technologies should be designed to help manage or mitigate the cognitive load limitations of the personnel in the convoy STeS.

## Convoy Organizational Structure

Additional automated truck capabilities will have an uneven impact on the collective tasks associated with convoy execution and will require reorganization of roles and responsibilities within the automated truck–enabled convoy. These changes will explicitly impact the organizational structure of the convoy. Current convoy structure is primarily driven by a triad of ratios between key force elements: PLS trucks, gun trucks, and personnel.

Furthermore, a particular impact of automated truck technology to the convoy organizational structure is in the direct reporting relationships and span of control.[2] Currently, almost all PLS trucks have communications capabilities so that truck crews can communicate directly with the CC or assistant convoy commander (ACC). Unmanned follower trucks will reduce the number of manned trucks with which the CC and ACC must coordinate.[3] However, each leader PLS in the automated truck–enabled convoy will have to provide oversight for and interface with the unmanned follower trucks. Although the span of control for the CC and ACC will be reduced, the span of control (and associated cognitive load) for leader PLS crews will significantly increase.

## Personnel

Inherent changes to convoy tasks, roles, and responsibilities and the increased cognitive loads will impact the proficiencies required for the personnel operating Army convoys. Because almost all personnel within the PLS convoy are from the 88M MOS, the introduction of automated trucks will significantly impact the 88M career progression over time.[4] The 88M MOS has four levels that represent progressively increasing levels of knowledge, skills, and abilities (KSAs) required to perform roles with increasing levels of responsibilities. Figure 5.2 depicts the four MOS levels with the roles generally performed. The roles listed in red text are ones that are likely to be significantly impacted by automated truck technology.

---

[2] *Span of control* refers to the number of functions, people, or things for which an individual or organization is responsible.

[3] The sensor suite of the applique kit will provide the CC and ACC greater situational awareness with access to truck locations, speeds, and visuals of the surrounding environment.

[4] Department of the Army, 2007. A United States Army MOS code is a grouping of duty positions requiring similar qualifications and the performance of closely related duties.

**Figure 5.2**
**88M MOS Pyramid with Key Positions by MOS**

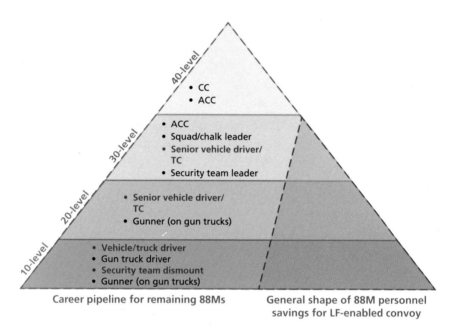

- **40-level**
  - CC
  - ACC

- **30-level**
  - ACC
  - Squad/chalk leader
  - Senior vehicle driver/ TC
  - Security team leader

- **20-level**
  - Senior vehicle driver/ TC
  - Gunner (on gun trucks)

- **10-level**
  - Vehicle/truck driver
  - Gun truck driver
  - Security team dismount
  - Gunner (on gun trucks)

Career pipeline for remaining 88Ms          General shape of 88M personnel savings for LF-enabled convoy

The reduction of personnel involved in convoy operations antici-pated with automated truck technology will not be proportional across MOS levels—there will be greater demands for more-senior drivers and fewer demands for entry-level drivers. As Figure 5.2 illustrates, the vast majority of the reductions will occur at the 10-level and 20-level 88M positions, with few or no reductions among 88M senior noncom-missioned officers (NCOs) at the 30 and 40 levels.[5] These changes will reduce the number of soldiers at risk but will not directly enable the generation of additional convoys and overall throughput increases. Additional shifts will require reorganization of existing transportation company force structure to increase the 30- and 40-level 88Ms rela-tive to the 10- and 20-level personnel. These demands will eventually change the fundamental structure of the 88M MOS career pyramid

---

[5]   The 10 level is private through specialist/corporal, the 20 level is a sergeant, the 30 level is a staff sergeant, and the 40 level is sergeant first class.

and possibly require alternative approaches to developing senior 88M personnel.

The vast majority of personnel impacts from introduction of automated trucks will occur for the 88M personnel, but other MOSs are likely to have impacts as well. For example, the 91B MOS (wheeled vehicle mechanic) is responsible for conducting maintenance, repair, and recovery within the PLS truck company.[6] These personnel will require new skills to diagnose and address maintenance faults on the new PLS with additive automated truck system components. Because of the technological sophistication of intended automated truck components, maintenance and repair of the automated truck–enabled PLS may require skills more consistent with the 94-series MOSs (electronic/missile maintenance).[7] However the Army adapts to these changes, automation will create attractive opportunities for soldiers to become automation technicians and operators.

---

[6]   91B MOS (wheeled vehicle mechanic) personnel are primarily responsible for supervising and performing maintenance and recovery operations on wheeled vehicles and associated items, as well as heavy-wheeled vehicles and select armored vehicles.

[7]   Alternative maintenance concepts could mitigate the impact to the 91B MOS.

# CHAPTER SIX

# Discussion and Recommendations

We examined the technical and tactical risks associated with automating Army trucks. Convoy operations are often required to cover extensive distances of unsecured routes. Convoys are particularly vulnerable to attack and ambush in noncontiguous and noncontinuous combat spaces that generally do not have secure rear areas. Convoy operations in the Iraq War and in Afghanistan further illustrated these vulnerabilities with extensive insurgent use of direct fires, coordinated ambushes, and many variations of IEDs. Almost all imaginable future scenarios include conventional and hybrid warfare aspects that will pose threats to convoys throughout the entire battlespace.

Our assessment of Army truck automation development, with the goal of reducing soldier casualties, compared the technical and tactical feasibility of two alternative concepts. In the MM employment concept, a leader truck provides the pathway for follower trucks being driven by an automated system. Though the follower trucks are driving themselves, there is still a soldier operator in the driver's seat monitoring the environment and taking back control of the truck in situations that the automated system is unable to manage. It is anticipated that only one soldier, instead of the current two, will need to be in the follower truck because the automated system will relieve the soldier of many of the driving tasks. This concept provides a 27-percent reduction in convoy personnel. The second employment concept, PU, is very similar to the MM employment concept; the major difference is that the follower trucks are completely unmanned. Because there will be situations that the automated system will be unable to manage, backup drivers traveling in more-secure gun trucks will need to dismount and drive the disabled automated follower truck. The PU concept provides a 37-percent reduction in convoy personnel.

Table 6.1 highlights our major findings.

Vehicle automation is a fast-evolving field. The commercial sector is making tremendous advancements. The Army will need to continue to monitor, and potentially partner with, the commercial industry. For these reasons, our research and recommendations focus on how the Army should proceed in the near term. Five major recommendations resulted from our assessment of the technical and tactical implications of automating Army trucks for convoy operations.

**Recommendation 1: Execute the MM employment concept as a necessary bridging strategy to achieve the full PU employment capability.** Current and near-term sensor and software technologies are not mature enough to successfully manage complex combat environments. The driving environment that automated Army trucks will need to successfully traverse is highly complex. Having the automated

**Table 6.1**
**Benefits of the MM Employment Approach to Address Most Major Concerns and Ensure Program Success**

| Concern Type | Benefits of MM Employment Concept (Versus PU Employment upon Fielding) |
|---|---|
| Technical concerns | • MM minimizes issues with maturing sensor and software capabilities to handle unexpected obstacles or actions<br>• Human operators are a key enabling factor to achieve the intended AV capability |
| Tactical concerns | • Creates larger trade space and flexibility for tactical commanders to tailor system employment based on dynamic threat and operating environment conditions<br>• Eliminates problematic and risky actions to return driver(s) to follower vehicles while in contact<br>• Provides ability to respond to nuanced tactical requirements (e.g., backing up) not fully accounted for by the initial capability |
| Acquisition challenges | • Provides a crossover period in which mileage can be accrued to validate and refine autonomous capabilities |
| Other concerns | • Provides an intermediate step to limit potential for overly optimistic force reductions prior to system reaching full maturity |

system follow the path of a truck being driven by a professional soldier does help reduce the complexity. However, the automated trucks are still required to sense and react to the driving environment, including actions by intelligent adversaries. The automated technologies, however, are sufficiently mature to handle many driving environments, and, if there is a soldier in the vehicle, the soldier can regain control of the truck in situations that the automated system is unable to handle. These events in which the soldier must regain control of the automated truck will provide opportunities to improve and mature the software and sensor capabilities over time. It is anticipated that the automated system will ultimately improve to the point at which the soldier operator could be removed from the vehicle. At this time, the PU employment concept will be achieved.

Pursuing the MM employment concept requires nearly all the same technology requirements as the PU employment concept, allowing the development program to continue to proceed under this current requirement. Furthermore, the MM employment concept allows the number of soldiers in the cab to be reduced from two to one, which maintains the primary motivation. However, this could be a major cultural adjustment for the Army because Army practice, for decades, has been to man each truck with two soldiers.

**Recommendation 2: Ensure a robust human factors design to mitigate employment risks inherent in the interim MM employment concept.** The major technical and tactical drawback to the MM employment concept is the human factors design, or HMI. There are three design and training aspects to the HMI: sensor perception communication, multimodal warning, and external communication. The soldier operator who monitors the automated system in the follower truck will be best suited for this task when she or he understands how the automated system perceives the world. This is commonly done by a display with a video feed of the driving view overlaid by what the automated system is sensing. There will be instances in which the sensor and software technology will struggle to sense the environment (e.g., glare from the sun). In these instances, the operator will need to be alerted. Studies show that robust training and multimodal alerting systems are required. Moreover, much of the nonverbal communication that occurs between other drivers and bystanders is lost when the vehicle is automated. This aspect of truck operations will need to be addressed to insure safe operations among other vehicles and people.

**Recommendation 3: Develop clear and practical technical requirements to reduce key development risks.** Several technical risks were uncovered in our assessment. Developing realistic, clear, and stable technical requirements is essential to avoid cost overruns, schedule slippages, and performance deficiencies. These technical risks are best managed early in the development process. For example, vehicles are vulnerable to cyberattack. The most effective way to secure a vehicle from such vulnerabilities is to include cybersecurity measures during the initial architecture development. As another example, establishing interoperability requirements early in development is crucial because it is anticipated that the Army will benefit from advances in sensor and software technology provided by the commercial automotive industry. Furthermore, automated vehicle path-following has been somewhat problematic. Because of this issue, it is important that the technology requirement is feasible and meets the tactical needs of the convoy operation. Lastly, obstacle detection and avoidance remains the most problematic technical capability to achieve. The requirement must properly balance associated cost and development time with the tactical needs. The MM employment concept eases the technical requirements for these last two areas; however, even under the MM employment concept, there will need to be an acceptable level of performance for path-following and obstacle detection.

**Recommendation 4: Use the MM approach to collect sustained user input for PU development and refinement.** User input is key to ensuring that the final product meets the intended purpose. Furthermore, user input, data collection, and data dissemination will be critical to the continued maturation of the automated system. The newness of AV technology and importance of HMI will make extensive and formalized user input a fundamental requirement for the success of Army automated truck implementation. User involvement will be fundamental for translating requirements into measurable technical specifications and assessing likely tactical implications of the technical decisions. Lastly, the automated truck development program should seek to leverage established Army training venues to demonstrate, validate, and build the confidence of key stakeholder groups (e.g., 88M NCOs).

**Recommendation 5: Prepare for the inevitable long-term technology, force structure, and personnel impacts resulting from automated truck emergence.** As described in Chapter Five, automated truck technology will significantly impact Army transportation operations over the long term and in many ways, including changes to force structure. Because it is reasonable to assume that pressure to leverage automated truck capability for force structure reductions will occur, there is the potential that force reduction decisions may be made prematurely. The Army must be prepared to respond to these pressures with accurate assessments of system capabilities and a full accounting of the range of tasks required of the units that own the trucks. Only such an analysis can identify the range of risks and benefits associated with force structure change proposals that will flow from the introduction of greater truck automation in convoy operations.

# Bibliography

Anderson, James M., Nidhi Kalra, Karlyn D. Stanley, Paul Sorensen, Constantine Samaras, and Tobi A. Oluwatola, *Autonomous Vehicle Technology: A Guide for Policymakers*, Santa Monica, Calif.: RAND Corporation, RR-443-2-RC, 2016. As of July 22, 2016:
http://www.rand.org/pubs/research_reports/RR443-2.html

Bikson, T. K., and J. D. Eveland, "Sociotechnical Reinvention: Implementation Dynamics and Collaboration Tools," *Information Communication and Society*, Vol. 1, No. 3, 1998, pp. 270–290.

Blanco, Myra, Jon Atwood, Holland M. Vasquez, Tammy E. Trimble, Vikki L. Fitchett, Josh Radlbeck, Gregory M. Fitch, Sheldon M. Russell, Charles A. Green, Brian Cullinane, and Justin F. Morgan, *Human Factors Evaluation of Level 2 and Level 3 Automated Driving Concepts*, Washington, D.C.: National Highway Traffic Safety Administration, Report No. DOT HS 812 182, August 2015.

Chen, Kan, Kenan Jarboe, and Janet Wolfe, "Long-Range Scenario Construction for Technology Assessment," *Technological Forecasting and Social Change*, Vol. 20, No. 1, August 1981, pp. 27–40.

Clegg, C. W., "Sociotechnical Principles for System Design," *Applied Ergonomics*, Vol. 31, No. 5, 2000, pp. 463–477.

Corbridge, C., and C. A. Cook, "Future Challenges for Function Allocation," *Proceedings of the Human Factors and Ergonomics Society Annual Meeting*, SAGE Publications, 1999.

Davis, James, and Ed Schoenherr, "Investigating the Performance of an Autonomous Driving Capability for Military Convoys," in Tadeusz Marek, Karwowski Waldemar, and Valerie Rice, eds., *Advances in Understanding Human Performance Neuroergonomics, Human Factors Design, and Special Populations*, Boca Raton, Fla.: CRC Press, 2010. As of January 11, 2017:
https://www.taylorfrancis.com/books/9781439835029

Davis, M. C., R. Challenger, D. N. W. Jayewardene, and C. W. Clegg, "Advancing Socio-Technical Systems Thinking: A Call for Bravery," *Applied Ergonomics*, Vol. 45, No. 2A, 2015, pp. 171–180.

Dearden, A., M. D. Harrison, and P. C. Wright, "Allocation of Function: Scenarios, Context and the Economics of Effort," *International Journal of Human-Computer Studies*, Vol. 52, No. 2, 2000, pp. 289–318.

Defense Science Board, *The Role of Autonomy in DoD Systems*, July 2012. As of January 11, 2017:
https://www.acq.osd.mil/dsb/reports/2010s/AutonomyReport.pdf

Department of the Army, "Pamphlet 611–21: Military Occupational Classification and Structure," 2007.

Diederichs, Frederik, Sven Bischoff, Harald Widlroither, Patrice Reilhac, Katharina Hottelart, and Julien Moizard, "New HMI Concept for an Intuitive Automated Driving Experience and Enhanced Transitions," September 2015. As of January 12, 2017:
https://www.auto-ui.org/15/p/workshops/4/
New%20HMI%20concept%20for%20an%20intuitive%20automated%20
driving%20experience%20and%20enhanced%20transitions.pdf

DoD—*See* U.S. Department of Defense.

Driverless Car Market Watch, "Forecasts," undated. As of January 11, 2017:
http://www.driverless-future.com/?page_id=384

DSB—*See* Defense Science Board.

Eason, Ken, Susan Harker, and Wendy Olphert, "Working with Users to Generate Organisational Requirements: The ORDIT Methodology," *ICL Systems Journal*, Vol. 11, No. 2, January 1997.

European Union, *European Truck Platooning Challenge 2016*, 2016.

Fraedrich, Eva, Sven Beiker, and Barbara Lenz, "Transition Pathways to Fully Automated Driving and Its Implications for the Sociotechnical System of Automobility," *European Journal of Futures Research*, Vol. 11, No. 3, 2015.

Fuld, R. B., "The Fiction of Function Allocation, Revisited," *International Journal of Human-Computer Studies*, Vol. 52, No. 2, 2000, pp. 217–233.

Google Auto LLC, *Google Self-Driving Car Testing Report on Disengagements of Autonomous Mode*, December 2015. As of January 11, 2017:
https://orfe.princeton.edu/~alaink/SmartDrivingCars/Papers/
GoogleDisengagementReport_2015.pdf

Greenberg, Andy, "Hackers Remotely Kill a Jeep on the Highway—with Me in It," *Wired*, July 21, 2015. As of January 12, 2017:
https://www.wired.com/2015/07/hackers-remotely-kill-jeep-highway/

Grote, Gudela, Cornelia Ryser, Toni Wäler, Anna Windischer, and Steffen Weik, "KOMPASS: A Method for Complementary Function Allocation in Automated Work Systems," *International Journal of Human-Computer Studies*, Vol. 52, No. 2, 2000, pp. 267–287.

Harker, S., and K. Eason, "The Use of Scenarios for Organisational Requirements Generation," Proceedings of the 32nd Annual Hawaii International Conference on Systems Sciences, Institute of Electrical and Electronics Engineers, 1999.

Heim, Scott, "Autonomous Mobility Applique System Joint Capability Technology Demonstration," U.S. Army TARDEC, April 1, 2015.

Hunter, Andrew Philip, Gregory Sanders, Jesse Ellman, and Kaitlyn Johnson, "Federal Research and Development Contract Trends and the Supporting Industrial Base, 2000–2015," Center for Strategic and International Studies, 2016.

Kalra, Nidhi, and Susan M. Paddock, *Driving to Safety: How Many Miles of Driving Would It Take to Demonstrate Autonomous Vehicle Reliability?* Santa Monica, Calif.: RAND Corporation, RR-1478-RC, 2016. As of October 29, 2019: https://www.rand.org/pubs/research_reports/RR1478.html

Killblane, Richard E., *Convoy Ambush Case Studies: Volume II—Iraq and Afghanistan*, Fort Lee, Va.: U.S. Army Transportation School, 2015. As of January 11, 2017: http://www.transportation.army.mil/history/publications/ 0336_0339_Ambush_Vol_2.pdf

Krzeszewski, John T., "Connecting the Phone to the Car—Protecting Personal Information and Securing the Interface," Delphi Automotive, presentation at Cyber Security Summit, Novi, Michigan, March 2016.

Leavitt, H. J., "Applied Organization Change in Industry: Structural, Technical, and Human Approaches," in S. Cooper, H. Leavitt, and K. Shelly, eds., *New Perspectives in Organizational Research*, New York: Wiley, 1964.

Lee, Connie, "Leader-Follower Scheduled for MDD in 2017," *Inside Defense*, November 1, 2016. As of January 11, 2017: https://insidedefense.com/daily-news/Leader-Follower-scheduled-mdd-2017

Litman, Todd, "Autonomous Vehicle Implementation Predictions; Implications for Transport Planning," Victoria Transport Policy Institute, August 2013.

Lyytinen, Kalle, and Mike Newman, "Explaining Information Systems Change: A Punctuated Socio-Technical Change Model," *European Journal of Information Systems*, Vol. 17, December 2008, pp. 589–613.

McCarty, Madeleine, Kelly Funkhouser, Jonathan Zadra, and Frank Drews, "Effects of Auditory Working Memory Tasks with Switching Between Autonomous and Manual Driving," *Proceedings of the Human Factors and Ergonomics Society Annual Meetings*, September 8, 2016.

Mercedes-Benz, "The Long-Haul Truck of the Future," undated. As of September 21, 2016:
https://www.mercedes-benz.com/en/mercedes-benz/innovation/the-long-haul-truck-of-the-future/

Miller, Robert B., *A Method for Man-Machine Task Analysis*, Wright-Patterson Air Force Base, Ohio: Wright Air Development Center, WADC Technical Report 53-137, June 1953. As of January 16, 2017:
www.dtic.mil/cgi-bin/GetTRDoc?AD=AD0015921

National Highway Traffic Safety Administration, *Comparison of Driver Braking Responses in a High Fidelity Driving Simulator and on a Test Track*, Springfield, Va.: National Technical Information Services, March 2002. As of January 12, 2017:
https://www.nhtsa.gov/DOT/NHTSA/NRD/Multimedia/PDFs/Human%20Factors/Driver%20Assistance/DOT%20HS%20809%20447.pdf

NHTSA—*See* National Highway Traffic Safety Administration.

Reinhold, M., "Protect from Cyber Threats—and Stay in Control," Robert Bosch LLC, presentation at Cyber Security Summit, Novi, Michigan, March 2016.

Rio Tinto Operations Centre, "Mine of the Future," Perth, Australia, 2014. As of January 11, 2017:
http://www.riotinto.com/australia/pilbara/mine-of-the-future-9603.aspx

Soetanto, Robby, Andrew R. J. Dainty, Andrew D. F. Price, and Jacqueline Glass, "Utilising Socio-Technical Systems Design Principles to Implement New ICT Systems," in D. J. Greenwood, ed., *Proceedings 19th Annual ARCOM Conference, 3–5 September 2003, Brighton, UK*, Vol. 2, 2003, pp. 695–704

Sommerville, Ian, "Socio-Technical Systems," *Software Engineering*, 7th ed., London: Pearson Education, 2011.

Strain, J., and K. Eason, "Exploring the Implications of Allocation of Function for Human Resource Management in the Royal Navy," *International Journal of Human-Computer Studies*, Vol. 52, No. 2, 2000, pp. 319–334.

Theisen, B., "Autonomous Ground Resupply (AGR) Science Technology Objective (STO)," Warren, Mich.: U.S. Army Tank Automotive Research, Development and Engineering Center, 2016.

U.S. Army, *Robotic and Autonomous Systems (RAS) Science and Technology Update*, Aberdeen Proving Ground, Md.: U.S. Army Research, Development and Engineering Command, 2016.

U.S. Army, *Soldiers Manual and Trainers Guide for 88M, Motor Transport Operator*, STP 55-88M14-SM-TG, 2013.

U.S. Department of Defense, *Unmanned Systems Roadmap 2007–2032*, Washington, D.C.: Government Printing Office, 2007. As of January 3, 2020:
https://www.globalsecurity.org/intell/library/reports/2007/dod-unmanned-systems-roadmap_2007-2032.pdf

Vyas, P., "Is the Car Lying to You?—Secure Connected Car Software Implementation," Harman International Industries, Incorporated, presentation at Cyber Security Summit, Novi, Michigan, March 2016.

Waterson, P. E., M.T. Older Gray, and C. W. Clegg, "A Sociotechnical Method for Designing Work Systems," *Human Factors*, Vol. 44, No. 3, 2002, pp. 376–391.

Weisbord, Martin R., *Productive Workplaces: Organizing and Managing for Dignity, Meaning, and Community*, San Francisco: Jossey-Bass, 1991.

Witten, Brian, "Building Comprehensive Security into Cars," Symantec, presentation at Cyber Security Summit, Novi, Michigan, March 2016.